Nuclear Radiation in Warfare

sipri

Stockholm International Peace Research Institute

SIPRI is an independent institute for research into problems of peace and conflict, especially those of disarmament and arms regulation. It was established in 1966 to commemorate Sweden's 150 years of unbroken peace.

The Institute is financed by the Swedish Parliament. The staff, the Governing Board and the Scientific Council are international. As a consultative body, the Scientific Council is not responsible for the views expressed in the publications of the Institute.

sipri

Stockholm International Peace Research Institute
Bergshamra, S-171 73 Solna, Sweden
Cable: Peaceresearch, Stockholm
Telephone: 08-55 97 00

Nuclear Radiation in Warfare

sipri

Stockholm International Peace Research Institute

Taylor & Francis Ltd
London
1981

First published 1981 by Taylor & Francis Ltd
4 John Street, London WC1N 2ET

Distributed in the United States of America and Canada by
Oelgeschlager, Gunn & Hain, Inc.,
1278 Massachusetts Avenue, Harvard Square,
Cambridge, MA 02138
and in Scandinavia by
Almqvist & Wiksell International,
26 Gamla Brogatan,
S-101 20 Stockholm, Sweden

British Library Cataloguing in Publication Data

Nuclear radiation in warfare.
1. Atomic warfare
I. Stockholm International Peace Research Institute
304.2 TD196.R3

ISBN 0-85066-217-6

Typeset by Performance Typesetting, Milton Keynes
Printed and bound in the United Kingdom by
Taylor & Francis (Printers) Ltd, Rankine Road,
Basingstoke, Hampshire RG24 0PR

Preface

The question of the consequences of nuclear wars has again become a much-discussed issue. The apparent emphasis in official nuclear policies on nuclear war fighting (counterforce) and discussions about the feasibility of a limited nuclear war are some of the reasons for renewed interest.

Some effects of nuclear weapons are quantifiable, others are unpredictable. It is essential that the latter are neither ignored nor forgotten. It is hoped that this book will help the discussion about the consequences of nuclear war. Hopefully its conclusion that no scientist can assure us that humankind could survive a nuclear world war involving the use of all or most of the nuclear weapons in current nuclear arsenals will always be taken into account by policy makers.

This book was written by Professor Joseph Rotblat (University of London) while a visiting scholar at SIPRI.

June 1981

Frank Barnaby
Director

Acknowledgements
The author wishes to express his thanks to Professors J.W. Boag and Patricia J. Lindop for helpful discussions, and to Billie Bielckus for editorial assistance.

Contents

Tables and figures

Tables

Figures

Units

Relationship between units

SI units (used in this book)		Other units	Relation between them
Energy:	joule (J)	electronvolt (eV)	1 J = 6.25×10^{18} eV
Pressure:	pascal (Pa)	atmosphere	1 Pa = 9.87×10^{-6} atmospheres
Temperature:	kelvin (K)	Centigrade (or Celsius) (°C)	1 K = 273.15 + °C

Prefixes of units

Multiples			Submultiples		
Factor	Prefix	Symbol for prefix	Factor	Prefix	Symbol for prefix
10^{3}	kilo	k	10^{-15}	femto	f
10^{6}	mega	M	10^{-12}	pico	p
10^{9}	giga	G	10^{-9}	nano	n
10^{12}	tera	T	10^{-6}	micro	μ
10^{15}	peta	P	10^{-3}	milli	m
10^{18}	exa	E			

Glossary

Actinides	The group of radioactive elements from atomic number 90 (thorium) to atomic number 103 (lawrencium).
Activity	A measure of the intensity of a radioactive source; it is equal to the number of nuclides disintegrating per second.
Alpha-particles	Helium-4 nuclei emitted spontaneously from some radioactive nuclides.
Atomic bomb (A-bomb)	A nuclear weapon in which the explosive energy is derived from fission only.
Atomic number	The place occupied by an element in the Periodic Table of Elements. It is determined by the number of protons in the nucleus of an atom.
Beta-particles (or beta-rays)	Fast moving electrons (or positrons) emitted spontaneously from the majority of radioactive nuclides.
Biological half-life	The time in which the amount of a (usually radioactive) substance contained in the body, or in a particular organ, is decreased to half as a result of natural biological processes.
Boosting	A method of enhancing the explosive yield of a fission weapon by neutrons produced in a small-scale thermonuclear reaction.
Chain reaction	A reaction that stimulates its own repetition. In a fission chain reaction a nuclide undergoing fission after the absorption of a neutron releases neutrons which can then cause further fission.
Circular error probable (CEP)	The radius of the circle with the target at its centre, within which a missile has a 50 per cent probability of landing.
Collective dose	A measure of the total burden on a population resulting from exposure to radiation; it is equal to the product of the dose and the number of persons exposed to it.
Cosmic radiation	Ionizing radiation which reaches the Earth from outer space; it contributes to the natural background radiation.
Counter-city strategy	A mode of employment of strategic nuclear weapons in which centres of population are the main targets.
Counterforce strategy	A mode of employment of strategic nuclear weapons in which the targets are the opponent's strategic nuclear forces.
Critical mass	The smallest mass of fissile material that will support a self-sustaining chain reaction under stated conditions. For an explosion to occur a mass greater than the critical (supercritical) is required.
Depleted uranium	Uranium in which the concentration of uranium-235 is lower than in naturally occurring uranium; it is practically entirely uranium-238.

Deuterium	An isotope of hydrogen of mass number 2; it occurs naturally in a proportion of one atom of deuterium to about 6 500 atoms of hydrogen of mass number 1.
Dose	A general term denoting the quantity of ionizing radiation absorbed by the body.
d–t reaction	The most important of the thermonuclear reactions in a fusion bomb; it occurs when a deuterium nucleus (deuteron) combines with a tritium nucleus (triton) to produce an alpha-particle and a neutron, with a large release of energy.
Electromagnetic spectrum	The electromagnetic radiations ranging (in order of increasing wavelengths) from gamma-rays, or X-rays, to ultraviolet, visible, infra-red, radar and radio waves.
Electron-capture	A rare mode of radioactive decay in which an orbital electron from the atom is captured by the nucleus; the result is the emission of X-rays.
Enrichment plant	A plant to produce substances containing isotopes in a different proportion than occurring naturally; usually refers to the production of uranium with a higher content of uranium-235.
Epicentre	The point on the ground vertically beneath an air explosion of a nuclear weapon; often called ground zero.
Equivalent megatonnage	A measure of the power of a nuclear weapon in terms of the mechanical effects it may produce; it is equal to the actual explosive yield (in megatons) raised to the power of two-thirds.
F-F-F bomb	A nuclear weapon in which energy is released in three stages: (1) fission—acting as a trigger; (2) fusion—occurring at the high temperature created in the first stage; and (3) fission—by the neutrons emitted at fusion, in a uranium tamper.
Fission	The splitting of a heavy nucleus into two approximately equal parts, accompanied by the release of energy and several neutrons.
Fission bomb	See atomic bomb.
Flexible response	A military doctrine, according to which a state may respond to an attack with means appropriate to the situation, for example, with conventional, tactical nuclear or strategic nuclear weapons. It is in contrast to the doctrine of mutual assured destruction (MAD).
Fluence	The intensity of radiation (number of particles, energy) falling on a unit of surface area.
Fratricide effect	The detonation of a second nuclear weapon on a target being inhibited by the effects (X-rays, thermal or blast waves) of the first weapon.

Fusion	The formation of a heavier nucleus from lighter ones, with an attendant release of energy; usually refers to the interaction of hydrogen nuclei to form helium.
Fusion bomb	A nuclear weapon in which the explosive energy is derived from fusion (apart from the fission trigger).
Gamma-rays	Electromagnetic radiation of high energy accompanying many nuclear reactions, such as fission, inelastic scattering or capture of neutrons, and radioactive decay.
Genotype	The genetic constitution of an individual.
GW(e)	Gigawatt electric—the power output of a reactor in the form of electricity.
Half-life	The time in which half of the number of nuclides in a given radioactive substance disintegrate.
Hydrogen bomb (H-bomb)	See fusion bomb.
Isotopes	Nuclides with the same atomic number.
Kelvin	The S.I. unit of temperature.
Laser beam for military purposes	A powerful beam of coherent monochromatic light.
LD-50	The dose of radiation required to kill 50 per cent of individuals in a population, within a specified period.
Light water reactor (LWR)	A reactor using slightly enriched uranium as fuel, and ordinary water both as moderator and coolant.
Linear energy transfer (LET)	The average amount of energy lost by an ionizing particle per unit path length.
Mass number	The number of nucleons (protons plus neutrons) in the nucleus of the given nuclide.
Neutron	An uncharged elementary particle with a mass slightly greater than that of the proton; the neutron is a constituent of nuclei of every atom heavier than hydrogen.
Neutron generator	In relation to fission bombs it is a device to ensure the presence of (at least) one neutron at the moment when the bomb is to be detonated.
Non-Proliferation Treaty (NPT)	The Non-Proliferation Treaty came into force in 1970; non-nuclear weapon states party to the Treaty agreed not to acquire or produce nuclear weapons.
Nuclide	Species of atom characterized by the number of protons and the number of neutrons in its nucleus. A nuclide is usually specified by giving the symbol of the element (which defines the atomic number) and the mass number, for example ^{235}U (or uranium-235).
Overpressure	The transient pressure exceeding the ambient pressure, in the blast wave from an explosion.
Photon	A quantum of energy of electromagnetic radiation; it is inversely proportional to its wavelength.

Positron	A particle identical with the electron, but with a positive electrical charge.
Prodromal syndrome	The early acute effects of exposure to radiation, sometimes called radiation sickness.
Proton	An elementary particle carrying a unit positive electrical charge; it is identical with the nucleus of hydrogen (of mass number 1), and is a constituent of the nuclei of all atoms.
Reactor	A system containing a controlled fission chain reaction; it is used to generate electricity, to produce plutonium, or for research.
Reprocessing plant	A plant in which spent reactor fuel is treated chemically to separate the plutonium and uranium from the fission products.
Slant distance	The distance from a given location on the surface of the Earth to the point where an explosion occurred.
Slow neutrons	Neutrons of low energy, of the same order as particles of a gas at a given temperature.
Somatic cells	All cells of the body other than germ cells.
Stratosphere	The layer of atmospheric air above the tropopause, in which the temperature changes very little with altitude.
Supercriticality	A state of a fission system when the amount of fissile material is greater than the critical mass under the existing conditions.
Tailings	The uranium ore left over after the extraction of the uranium from it.
Tamper	A material used to reflect the neutrons in a bomb assembly as well as to provide greater inertia.
Ternary fission	A rare fission process in which a nucleus splits into three parts.
Thermal neutrons	Low energy neutrons, in thermal equilibrium with the surrounding medium.
Thermal reactor	A reactor in which thermal neutrons are used to produce fission.
Thermonuclear bomb	A weapon in which part of the explosive energy is derived from fusion reactions.
Thermonuclear reaction	A fusion process brought about by very high temperatures.
TNT	A chemical explosive, trinitrotoluene; used as a measure of the energy released in the detonation of nuclear weapons.
Tritium	A radioactive isotope of hydrogen of mass number 3.
Tropopause	The boundary between the troposphere and the stratosphere.

Troposphere	The region of the atmosphere immediately above the Earth's surface, in which the temperature falls with increasing altitude.
Weathering	The natural reduction of the concentration of fall-out radioactivity by the action of rain and wind.
X-rays	Electromagnetic radiation, identical with gamma-rays, but produced in processes outside the atomic nucleus.

1. Introduction

The two atom bombs dropped on Japanese cities brought World War II to an end. Since then there have been over 100 wars (estimates vary between 100 and 300, depending on the criteria used to define war) in various parts of the world with many millions of casualties. But nuclear weapons have not been used again, although the nuclear arms race has continued unabated. The majority of people in the world today were born after the Hiroshima and Nagasaki bombings, and there is a tendency among them to believe that we have learned how to live with the bomb, and that the nuclear arms race need not lead to a nuclear war. This is wishful thinking. A nuclear war between the two great powers has so far been avoided—although we have come close to it on several occasions—because of the mutual balance of terror which makes it improbable for one side to be able to attack the other without the risk of incurring instant retaliation. However, any arms race is intrinsically unstable, and unless brought to a complete halt it is bound to lead to a catastrophe. A great variety of circumstances, created intentionally or arising accidentally, may trigger the use of nuclear weapons, and although the probability of the occurrence of any one triggering event is very small, it is certainly not zero. When taken cumulatively and over a period of time, these circumstances make the outbreak of a nuclear war highly probable.

The recent adoption by the United States government of the counterforce strategy is believed by many to have increased considerably the probability of a nuclear war. In this strategy the targets are military installations, rather than centres of population. The doctrine of flexible response, of which the counterforce strategy is one aspect, is based on the fallacious assumption that a nuclear war could be kept limited in scope and/or terrain. The belief that a nuclear war could be won at a relatively low cost in casualties may itself make the outbreak of such a war more likely.

A further increase of the probability of a nuclear war arises from the fact that the nuclear arms race is not confined to only two states. Although the possession of nuclear weapons by France and the United Kingdom does not make a significant difference, since even France is effectively tied up with the United States' policy, China must be considered as a third power, and this makes the balance of terror much more precarious. By exploding a nuclear device, purportedly for peaceful purposes, India has made other nations sufficiently apprehensive to make attempts to acquire nuclear weapons. The acquisition of the necessary technology and materials through the development of peaceful nuclear energy will accelerate this process of horizontal proliferation. Considering the instability of regimes in many countries, and the many causes of conflict which have resulted in some three to eight wars per year during the last few decades, the chance of nuclear weapons being used in such wars will greatly increase in the course of time.

The difference between a nuclear war and one fought with conventional weapons is both quantitative and qualitative. The destructive power per unit weight of the bomb is immensely greater in nuclear weapons. A single nuclear bomb can have an explosive yield greater than that of the total of all the explosives ever used in wars since gunpowder was invented. The qualitative difference which makes nuclear weapons unique is that, in addition to causing loss of life by mechanical blast, or by burns from the heat of the fire-ball, nuclear weapons have a third killer—radiation. Moreover, and unlike the other two agents of death, the lethal action of radiation can stretch well beyond the war theatre and continue long after the war has ended, into future generations.

There has been a tendency to minimize the contribution made by radiation to the total destruction and to describe it as insignificant compared with casualties from blast and heat effects. Many of the published studies of the consequences of nuclear warfare put the emphasis on the shock-wave phenomena and the blast effects accompanying the burst of a nuclear bomb. Even the explosive yield of a nuclear weapon is often expressed in terms of 'equivalent megatonnage' (actual yield to the power of two-thirds) which is a measure of the mechanical damage caused by the bomb. There are two reasons for this. First, long experience with conventional explosives makes it possible to calculate the effects of the blast wave fairly accurately, whereas the thermal and radiation effects are much less easy to assess. Thus, for 'mutual assured destruction' (MAD), which was the basis of the deterrence doctrine, the mechanical effects were primarily used in calculating the damage. Second, the counterforce strategy aims at the destruction of missile bases, for which the mechanical effect of the explosions is the deciding factor.

It has also been argued that for large bombs the areas within which the lethal effects of radiation occur are well within the lethal areas of thermal and blast effects. But this argument ignores the lethal effects of early fall-out. Indeed, in a counterforce attack, believed to be the most likely scenario for a nuclear war, the casualties due to radiation will probably outnumber those caused by blast and heat. This arises from the fact that the optimum condition for the destruction of missiles in hardened silos, that is, a surface burst, is also the condition for maximum early fall-out. In other types of nuclear warfare too, the areas covered by fall-out far exceed those in which the immediate effects occur. Nuclear radiation must therefore be considered as a major contributor to the casualty toll of a nuclear war.

However, despite numerous tests of nuclear weapons in the atmosphere (before the Partial Test Ban Treaty) and the study of the Japanese bomb survivors, it is still very difficult to make a quantitative estimate of the damage to man caused by radiations from nuclear weapons.

First, there is a wide range of methods for the exploitation of the lethal effect of radiation in nuclear weapons. At one end of the range, bombs can be devised such that there would be relatively little harm from blast and heat and almost all the damage would come from radiation, as with the so-called neutron bomb. At the other extreme, bombs can be detonated in such a way that there would be only global fall-out, and those receiving lethal doses of the initial radiation would in any case be killed by the blast and heat. In between there is a continuous spectrum of ways of employing nuclear weapons and of ways in which radiation could contribute to the total casualties. How weapons are likely to be employed in war depends on many factors, and the possibility cannot be excluded that maximization of the radiation killing effect would be aimed at.

Second, for a given method of employing a nuclear bomb, the extent of the fall-out peril to the population depends on many variables, such as the meteorological conditions in the period following the explosion, the speed and direction of winds, the time of day, the season of the year, and the demographic and geographic features of the terrain. It also depends on factors which the side attacked can control to a certain extent, such as adequate warning, availability and utilization of shelters, and efficient evacuation schemes. For these reasons, estimates of the radiation doses received and the number of people affected vary enormously, even for a war of postulated mode and scale.

Finally, even when the radiation dose can be estimated there is considerable uncertainty about its damaging effect on humans. Both at high dose levels, which may cause death within a short time after exposure, and at lower dose levels, which give rise to long-term effects such as cancer or genetic damage, not enough is known to predict the outcome accurately. Depending on the assumptions made, the eventual mortality may differ by a large factor. This applies also to the estimates of casualties which may result from global fall-out in countries remote from the combat areas.

Another serious defect in estimating the consequences of a nuclear war is that the casualties resulting from the three main factors—blast, heat and radiation—are usually calculated separately, as if each were present alone. But the superposition of all three, together with the economic, social and psychological traumata of a nuclear war, could make an enormous difference to the number of casualties. In particular, many deaths would result from injuries from which recovery would normally be complete.

All these unknown factors make it impossible to calculate accurately the extent of the loss in human lives and health, of present and future generations, which might result from the radiation effects of a nuclear war. Nevertheless, because of the uniqueness and all-importance of nuclear radiation in warfare, it is necessary at least to understand the bases on which such calculations can be made and the degree of uncertainty attached to them.

2. Digest of nuclear weaponry

2.1. Characteristics of nuclear weapons

2.1.1. Fission bombs

The basic nuclear weapon is the fission bomb, popularly known as the A-bomb. The energy is derived from the break-up, or fission, of the nuclides of very heavy atoms, such as uranium or plutonium, by the impact of a neutron. The three nuclides of interest are uranium-233, uranium-235 and plutonium-239, but only the last two are known to have been used in nuclear weapons. The main characteristics of the fission process are: (*a*) the large amount of energy released at each fission, 32 pJ (200 MeV) as compared with a millionth of a picojoule per atom in chemical reactions; (*b*) the radioactivity of the nuclides resulting from the splitting of the original nuclide; and (*c*) the release of several neutrons. It is this last property that makes the process feasible on a practical scale, as these neutrons can be used to produce fission in other nuclides, thus starting a divergent chain reaction.

For such a chain reaction to take place it is essential to ensure that the neutrons emitted at fission will not escape from the fissile material or be otherwise lost before hitting another fissile nuclide. The greater the mass of the material the smaller is the probability of a neutron escaping. There exists a critical mass, the smallest amount of fissile material in which a self-sustained chain reaction can take place. Apart from the nuclear properties of the material, the critical mass depends on several other factors. One is density: the higher the density of the material, the shorter is the average distance travelled by a neutron before causing another fission, and therefore the smaller the critical mass. Another factor is the purity of the material: the smaller the admixture of other materials, the smaller is the probability that neutrons will interact with them instead of causing fission, and therefore the smaller the critical mass. A third factor is the physical surroundings: if the fissile material is placed in a medium which reflects neutrons but does not absorb them, some of the neutrons which would otherwise have escaped can be utilized in the chain reaction, thus making the critical mass smaller. With a good reflector the critical mass can be reduced severalfold. Table 1 gives the critical masses for uranium and plutonium of the purity and density used in weapons, when surrounded by different reflectors. As is seen, beryllium is a better reflector than uranium, but a combination of both materials may be used for reasons described below.

Table 1. Critical masses (for materials of normal density)

Nuclide	Reflector	Thickness (cm)	Critical mass (kg)
Uranium-235	None		49
Uranium-235	Beryllium	10	14
Uranium-235	Natural uranium	10	18
Plutonium-239	None		12.5
Plutonium-239	Beryllium	5.2	5.4
Plutonium-239	Beryllium	32	2.5
Plutonium-239	Natural uranium	5	6.4
Plutonium-239	Natural uranium	24	4.4

Source: Argonne National Laboratory, *Reactor Physics Constants* (US AEA, 1963), pp. 582–84.

The neutrons emitted at fission have a very high velocity (about 2×10^7 metres per second on average) and they travel a few centimetres between collisions, so that it takes less than a hundred-millionth of a second (10^{-8} s) to hit another nuclide and produce fission. If more than one neutron is available after each fission to produce further fission, the number of neutrons will increase exponentially and so will the energy released. In the actual set-up for the atom bomb it would take about 56 generations of fission to multiply the neutron population so that 3.2×10^{24} fissions would take place and the total energy released immediately (29 pJ per fission) would be 92 TJ. Such an amount of energy is released in the detonation of 22 000 tons of the ordinary explosive TNT. Thus, the bomb would have an explosive yield of 22 kt of TNT; this was the size of the Nagasaki bomb. The actual amount of fissile material used up was 1.25 kg.

When a comparison is made with ordinary explosives it should be noted that the actual amount of energy released at the detonation of a given amount of TNT varies somewhat. For this reason it is conventional to refer to a 1-kt bomb as one in which 1.45×10^{23} fissions release immediately an energy of 4.18 TJ (10^{12} calories in the old units); the mass that underwent fission is then 57 g.

The required 56 generations in the above example will occur in less than a microsecond, but it is important to note that 99 per cent of the energy is released in the last five generations, which occur in about 0.05 microseconds. In other words, a difference in timing of a small fraction of a microsecond can make an enormous difference in the amount of energy released. This has a vital bearing on the method of assembly of the bomb.

As soon as an amount of material larger than the critical value is put together an explosion takes place, because only one neutron is needed to start the chain reaction and from cosmic radiation alone there will be a few neutrons hitting the fissile mass every second. For this reason the bomb consists of two or more pieces of the fissile material, each smaller than the critical mass, and these are brought together at the moment of the explosion. But unless this assembly takes place extremely fast, the energy released in the earlier fissions will disperse the material before the last few vital generations have occurred, and the explosive power will be quite small.

Two methods of rapid assembly are known to have been used. In one, the gun method, the fissile material is in two parts, placed at opposite ends of the barrel of a gun. At the required moment the two pieces are brought together by shooting one into the other, using an ordinary explosive. In the second method, based on the implosion technique, a sphere of the fissile material smaller than the critical mass is surrounded by segments of suitably-shaped plastic chemical explosives. At the required moment all these charges are simultaneously exploded, pushing the fissile material inwards and compressing it to a high density, thus reaching supercriticality. It should be noted that the critical masses given in table 1 are for materials at normal densities. At the much higher densities momentarily achieved in the implosion technique, an amount of material several times smaller is sufficient to achieve criticality.

To initiate a chain reaction, a neutron source is necessary to ensure that a neutron will be present at the right moment. Here again, timing is of vital importance; the neutrons must not arrive too soon as they could cause a pre-detonation. For this reason the neutron source is assembled at the same time as

Figure 1. Configuration of components of a fission bomb. A—initiator (neutron source or generator), B—fissile core (plutonium and U-235), C—tamper cum reflector (uranium plus beryllium), D—high explosive lens (shaped plastic charge), E—detonator.

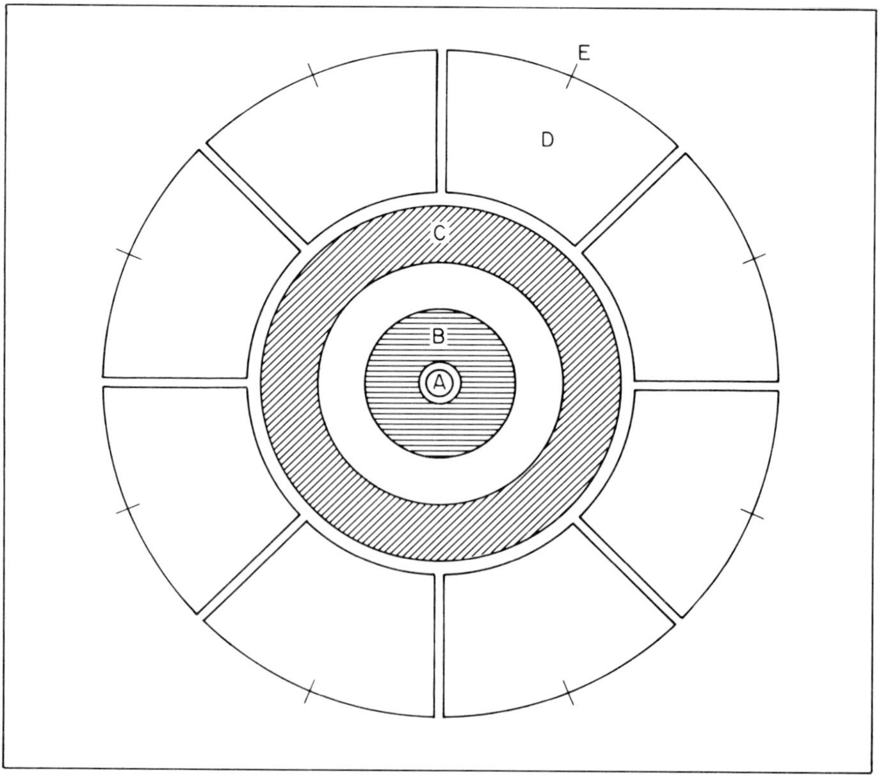

the fissile material. This can be achieved, for example, by bringing together beryllium or boron with an alpha-emitting radioactive material (the alpha-particles hitting the nuclei of these elements cause the emission of neutrons), or by using a neutron generator.

The explosive yield of the bomb depends on the amount of fissile material that undergoes fission before the whole assembly is blown apart by the force of the explosion. The neutron reflector plays an important role in this respect by acting as a tamper, to hold the material together just a little longer to enable the last vital generations of fissions to take place. For this reason, a reflector of high inertia, that is, made of an element of high density, is preferable. A system consisting of two materials, beryllium and uranium, can provide the optimum combination of a reflector and tamper. Figure 1 shows diagrammatically a possible arrangement of the different components in a plutonium fission bomb, using the implosion technique.

The existence of a critical mass limits the explosive power that can be obtained with a fission bomb, since it restricts the amount of fissile material that can be used. By refining the implosion technique it is possible to assemble a larger amount and utilize it more efficiently, but there is a limit to the improvements which technology can accomplish in this respect. An explosive yield of about 500 kt is considered to be the maximum achievable.

At the other extreme, by controlling precisely the amount of fissile material in the bomb and the method of assembly, much lower explosive yields can be achieved. Such bombs, used as shells in artillery guns, can have explosive powers as low as 10 tons of TNT, thus bridging the gap between nuclear and conventional weapons as far as explosive power is concerned.

2.1.2. Fusion bombs

Fusion bombs, also called thermonuclear or H-bombs, derive their energy from a process which is the reverse of that of fission, namely the formation of heavier nuclides from lighter ones. The only element which has so far been used for this purpose is hydrogen. Four types of fusion reaction between the heavier isotopes of hydrogen, deuterium (^2D) and tritium (^3T), are relevant; they are listed in table 2. It will be noted that three of these reactions are accompanied

Table 2. Fusion reactions

Reaction	Energy released (pJ)
1. $^2D + {}^2D = {}^3He + {}^1n$	0.15
2. $^2D + {}^2D = {}^3T + {}^1H$	0.64
3. $^3T + {}^2D = {}^4He + {}^1n$	2.82
4. $^3T + {}^3T = {}^4He + {}^1n + {}^1n$	1.81

by the emission of neutrons. The last column in the table gives the amount of energy released in each reaction. This is at least an order of magnitude smaller than that released at fission. But, considering the masses of the nuclides involved, it is seen that, per unit of mass, fusion reactions are on the average about three times more efficient than fission reactions.

Unlike fission, where one neutron is sufficient to start the chain reaction, a self-contained fusion reaction cannot start until the component nuclei have sufficiently high energy to enable them to overcome the repulsive force between them (this force results from the positive electric charges of the nuclei). The necessary energy can be supplied in various ways, but the method used in the bomb is to increase the energy of the particles by raising the temperature of the material. Hence the name thermonuclear reaction. It turns out that even for the deuterium–tritium (d–t) reaction (third reaction in table 2), which is the easiest to achieve, a temperature of nearly 100 million kelvin is required to make it self-sustained.

In principle, such a temperature could be created on a small scale without nuclear means, for example by directing a very powerful laser beam onto an appropriate mixture of deuterium and tritium. But although research into this method has been going on for some time, no success has yet been reported, and the only way to produce the necessary high temperature is the A-bomb in which such high temperatures are achieved at the instant of the explosion. Thus, every H-bomb consists of at least two stages: a fission bomb, which acts as a trigger, and a fusion mixture, which is ignited in the heat created by the trigger.

The direct use of tritium, which is a radioactive gas with a relatively short half-life (12.3 years), would be cumbersome, but this difficulty can be overcome by using a solid compound lithium deuteride (LiD) in which the lithium consists mainly of the isotope ^{6}Li. The neutrons produced at fission transform this element into helium and tritium, in the reaction

$$^{6}\text{Li} + {}^{1}\text{n} = {}^{4}\text{He} + {}^{3}\text{T}.$$

The tritium thus created interacts with the deuterium (according to reaction 3 in table 2), releasing fusion energy. The A-bomb trigger therefore not only creates the high temperature, but also helps to produce the material for the thermonuclear reaction.

Since there is no critical mass for the fusion process there is, in principle, no limit to the total energy released or to the explosive power of the H-bomb. However, in practice the utilization of a large amount of the fusion mixture (deuterium with tritium and/or lithium deuteride) would be impossible if the latter were simply placed around the fission trigger. The reason is that the very high temperature and pressure created by the trigger would blow the whole system apart before any but a small fraction of the fusion material became ignited. As in the A-bomb, the use of a heavy metal as a tamper would help somewhat in this respect, but not enough to give a large yield. The problem is to deliver sufficient energy to the bulk of the fusion material to start the thermonuclear reaction in a time which is short compared with the time in which the explosion occurs; indeed, the energy would have to be delivered at nearly the speed of light to achieve this.

The solution to the problem lies in the fact that, at the very high temperature of the fission trigger, most of the energy is emitted in the form of X-rays. These X-rays, travelling at the speed of light, radiate out from the centre and, on reaching the tamper, are absorbed in it and then immediately re-emitted in the form of softer X-rays. By an appropriate configuration of the trigger and the fusion material it is possible to ensure that the X-rays reach the latter almost instantaneously. If the fusion material is subdivided into small portions, each surrounded with a thin absorber made of a heavy metal, the bulk of the fusion material will simultaneously receive enough energy to start the thermonuclear reaction before the explosion disperses the whole assembly.

By the use of such a technique it is possible to achieve a very large explosive yield. Indeed, an H-bomb with an explosive power of about 60 Mt (60 million tons of TNT), that is, nearly 3 000 times more powerful than the Nagasaki bomb, was detonated by the Soviet Union in a test explosion in 1961, and there is no reason why higher yields could not be obtained, if desired.

Large bombs are economically advantageous, since the cost per megaton decreases rapidly as the megatonnage increases. For example, a 100-Mt bomb might cost only twice as much as a 100-kt bomb. Large bombs are also feasible from the point of view of weight, since for large H-bombs each kilogram of the thermonuclear mixture increases the explosive yield by about 10 kt, and the material for a 100-Mt bomb, for example, would weigh some 15 tons [1].

The practical limitation to the explosive yield of the H-bomb is due to another factor—its overkill capacity. Even the largest population centre could be destroyed by one 10-Mt bomb; with larger bombs, most of the energy would be wasted. Much less energy is needed to destroy a missile base; for this reason, most of the United States' warheads, and increasingly also those of the Soviet Union, intended to destroy the adversary's land-based missiles, have yields well below 1 Mt. However, when more than one low-yield bomb is considered to be necessary for a single target, the use of larger bombs may be more expedient, due to the 'fratricide effect' in which the disturbance produced by the explosion of one bomb inhibits the detonation of another bomb in quick succession.

It will be noted from table 2 that apart from tritium, which represents a relatively small radiological hazard, no radioactive materials are used or produced in the thermonuclear reactions. Therefore, in large H-bombs, where the energy contributed by the fission trigger is only a small fraction of the total, the damaging action of radioactivity is relatively small. A weapon which produces little radioactivity is referred to as 'clean'. From this point of view the H-bomb discussed so far can be described as a 'clean' weapon.

2.1.3. Booster bombs

If in the fission bomb (figure 1) the plutonium were in the form of a hollow sphere and a small amount of fusion material were placed in its centre, the compression and heating caused by the implosion would result in a small-scale thermonuclear reaction. The large number of neutrons emitted in this reaction would hit the plutonium and produce fissions, in addition to those from the original chain reaction. In this way the total fission yield could be greatly increased. In the jargon of bomb-makers this method is referred to as 'boosting'. Booster bombs were tested even before the H-bomb, described in the previous section, was developed.

2.1.4. Fission-fusion-fission (F-F-F) bombs

If the high inertia tamper in the H-bomb, needed to prevent premature dispersion, is made of uranium, an additional effect is obtained: the neutrons emitted in the thermonuclear reactions cause fission in the uranium, releasing more energy and thus adding very considerably to the total destructive power. The bulk of uranium, that is, the isotope uranium-238, does not undergo fission with slow neutrons. Therefore, for power production in nuclear reactors based on slow neutrons, uranium-238 is useless, except indirectly through its transformation into plutonium. But the neutrons released in the thermonuclear reactions are very fast and can induce fission in uranium-238. Natural uranium, or depleted uranium (uranium with a greatly reduced concentration of uranium-235, large quantities of which have accumulated as the tailings in enrichment plants), can be used as a tamper in the H-bomb to make it more efficient.

A bomb with such a uranium tamper releases energy in three stages: first through fission in the trigger, second through fusion, and third through fission in the tamper. For this reason it is called a fission-fusion-fission bomb. In the third stage huge quantities of radioactive substances are produced, just as in the first stage, but on a much larger scale, since many more neutrons result from the second stage than from the first. In such a bomb a high proportion of the total explosive yield is contributed by fission, this proportion depending on the construction of the bomb. Since for the same payload of the weapon a much greater explosive power is obtained if the third stage is utilized, it is reasonable to assume that in reality a three-stage bomb would generally be used. Indeed, in textbooks of nuclear weaponry a thermonuclear weapon is assumed to be an F-F-F bomb in which half of the energy released is due to fission and half to fusion.

In the conventional nomenclature such a bomb has to be considered as very 'dirty', since the amount of radioactivity produced can be thousands of times greater than in a fission bomb. Huge quantities of plutonium are also produced as a result of the explosion, since the capture of neutrons by uranium sometimes leads to its transformation into plutonium instead of to fission.

2.1.5. 'Salting'

Even 'dirtier' bombs could be devised by mixing the thermonuclear material with elements which have a high probability of being transformed into radioactive forms by the action of fast neutrons. Such a process is called 'salting' and its purpose would be to increase the hazard arising from the radioactivity without a corresponding increase in blast and heat effects. If a material were chosen which combines an optimum half-life with the emission of high-energy radiation (e.g., cobalt-60) an awesome radiological weapon might result.

2.1.6. Enhanced radiation weapons (neutron bombs)

At the other end of the spectrum of H-bombs is the enhanced radiation weapon, which has a very low yield, about 1 kt, but which is designed to minimize the effects of blast and heat and to maximize the effects of the neutrons. Hence the name neutron bomb.

In nuclear reactions the energy released appears initially as the kinetic energy of the nuclides formed in the reaction; under conditions applicable here, this kinetic energy is shared in inverse proportion to the masses of the nuclides. As shown in table 2, in the main fusion reaction (reaction 3) the neutron has one-fifth of the total mass; it therefore acquires four-fifths of the total energy. If the assembly of the bomb were such that little extraneous material were present to absorb the neutrons, 80 per cent of the total energy released would be carried away by the neutrons and only 20 per cent would be left to produce the usual blast and heat effects. In practice, the proportion of the neutron energy would be smaller, since a fission assembly is still needed to trigger the fusion reaction. But even so, the most damaging potential of this bomb would be due to the action of neutrons on living organisms. The main purpose of the neutron bomb is to kill the crews of tanks without causing much collateral damage.

Although by conventional standards the neutron bomb would be described as a 'clean' weapon, since very little radioactivity would result from it, the lethal radiation effect per unit of explosive yield would be greater than in the very 'dirty' weapons.

2.1.7. Distribution of energy

The great variety of types and designs of nuclear weapons means that the different forms of energy which cause the damaging action—blast, heat and radiation—may have significantly different distributions, even if the neutron bomb is disregarded. The distribution of energy also depends markedly on the medium in which the weapon is exploded, whether underwater, underground or in the air, and—in the latter case—on the altitude of the explosion. From the point of view of radiation effects, the most important are explosions on the surface or at low altitudes (although global fall-out occurs even with high-altitude bursts). For fission bombs, about 50 per cent of the total energy goes into the blast effect, 35 per cent is in the heat effect, 5 per cent appears in the form of initial radiation, and 10 per cent appears as residual radiation. In a thermonuclear bomb, with half its yield coming from fission, the residual radiation carries 5 per cent of the total energy. Since residual radiation is defined as that emitted starting one minute after the explosion (see chapter 4), its contribution is not generally included in the total energy release; the yield of weapons in terms of kilotons or megatons is, by convention, expressed as the energy released within the first minute.

2.2. *Effects of nuclear weapons other than ionizing radiation*

Since our main concern is with the effects of radiation, the damaging action of the two other main contributors, heat and blast, is discussed only briefly, with emphasis on their relevance to the radiation effects. The electromagnetic pulse effect is also mentioned in the same connection. The damaging action of nuclear weapons resulting from the emission of ionizing radiation is discussed in chapter 4.

2.2.1. The fire-ball and early fall-out

At the instant of the detonation of an A-bomb the temperature of the fission products and other parts of the weapon rises to tens of millions kelvin, and the pressure up to terapascals (millions of atmospheres). A body at an elevated temperature emits electromagnetic radiation; at the high temperature of the bomb the radiation is of short wavelength (of the order of a nanometre), and lies in the soft X-ray region of the electromagnetic spectrum. About 75 per cent of the total energy released may be carried by these X-rays. At this stage there is hardly any radiation in the visible region. However, soft X-rays are easily absorbed in the surrounding air. For explosions at not too high altitudes (less than 30 000 metres) the absorption is practically complete within a few metres. The air, being at a lower temperature, radiates energy of longer wavelength, which in turn is absorbed and re-radiated, so that there is a gradual transition from X-rays to the emission of ultraviolet, visible and infra-red radiation. This process leads to the formation of a luminous mass of air, the so-called fire-ball. Immediately after its formation, the fire-ball begins to grow and to rise in the air; as its mass increases, its temperature drops further. After about a minute the temperature has dropped sufficiently for the fire-ball not to emit visible radiation any longer. The size of the fire-ball depends on the yield of the weapon; a tenfold increase of the yield increases the size of the fire-ball by a factor of 2.5. For a 1-Mt bomb the maximum radius of the fire-ball is about 1 200 metres.

For explosions above ground the altitude of the burst and the size of the fire-ball have an important bearing on the magnitude of the early fall-out. If the fire-ball touches the ground, the soil and other materials are vaporized and taken up with the fire-ball. The strong after-winds cause huge quantities of dirt and debris to be sucked up. They mix with the radioactive fragments of the bomb and form particles of various sizes which move upwards as well as spreading out. Later they begin to fall to the ground under the force of gravity, at rates and distances depending on the size of the particles and the velocity of the wind. This deposition of radioactivity constitutes the early, or local, fall-out. On the other hand, if the bomb is exploded at such a height that the fire-ball never touches the ground there is much less or even no local fall-out (but there will be global, or delayed, fall-out). The maximum height of the burst for which appreciable local fall-out will occur is plotted in figure 2 as a function of the yield of the bomb.

Thus, for a 1-kt bomb there will be no local fall-out if the detonation is at a height greater than 55 metres above ground; for a 1-Mt bomb the relevant height is 860 metres. As is seen from the figure, for every tenfold increase in explosive yield, the maximum height for the occurrence of early fall-out increases by a factor of 2.5.

Figure 2. Maximum height of burst for occurrence of appreciable early fall-out

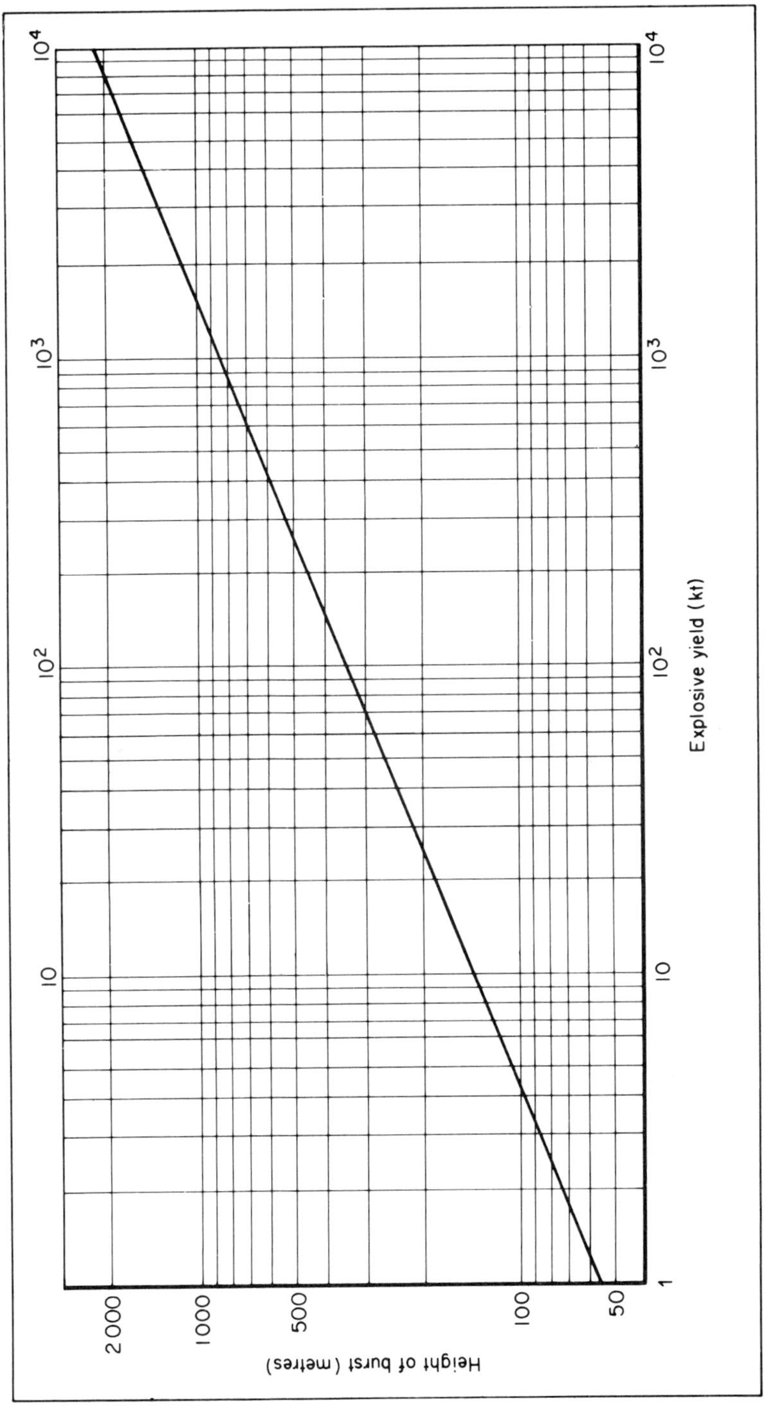

A detonation is usually designated as an air burst if it occurs at a height greater than that in figure 2 for the given yield. Detonations at lower heights are designated as surface bursts. However, it should be noted that small amounts of early fall-out may occur even with air bursts.

It should be borne in mind that the values in figure 2, as well as in subsequent figures which contain quantitative data about effects of nuclear bombs, have considerable errors attached to them; these errors arise partly from lack of sufficient knowledge about the phenomena discussed, and partly from the dependence of the magnitude of a given effect on a number of parameters which cannot be specified in detail in each case.

2.2.2. Thermal radiation

The electromagnetic waves radiated from the fire-ball due to its high temperature are termed thermal radiation. An analysis of the mechanism of the formation of the fire-ball shows that the thermal radiation is emitted in two pulses: first, a short pulse, of some milliseconds duration, and subsequently a longer pulse, of the order of seconds; the duration of the second pulse increases with increasing explosive yield of the bomb. The first pulse consists mainly of ultraviolet radiation; it carries only a tiny fraction of the total energy radiated, but it may damage the eyes of a person who happens to look in the direction of the explosion. The second pulse carries the bulk of the thermal energy and is responsible for the heat effects of the bomb.

The thermal radiation from this pulse can cause fatal or severe burns, and start fires, over a large area. Its range depends on the explosive yield of the bomb and atmospheric conditions; with fairly good visibility, the heat wave can cause casualties at much greater distances than for the other immediate effects—blast and initial ionizing radiations.

The distance at which a given amount of thermal radiation energy falls on a unit area should vary as the square root of the explosive yield, but owing to absorption and scattering by particles of air, the distance increases more slowly. Furthermore, the injury produced by exposure to a given amount of thermal energy depends on the duration of the thermal pulse: the longer its duration, the more thermal energy is required to produce a given degree of burn. Since with increasing yield the heat pulse lasts longer, this also results in a slower increase with explosive yield of the distance at which a given degree of thermal injury is produced.

Figure 3 shows the slant distance at which there is a 50 per cent probability of third-degree burns being suffered by persons caught in the path of the thermal radiation, as a function of the explosive yield. The curve applies to explosions at low altitudes and at a visibility of about 20 km. As is seen, for a 10-kt bomb the distance is nearly 2 km, while for a 10-Mt bomb it is more than 22 km. Persons exposed to third-degree burns can survive if they receive medical treatment. Under war conditions this is unlikely to be the case, and figure 3 should therefore be considered as presenting a measure of the lethal distances for thermal effects.

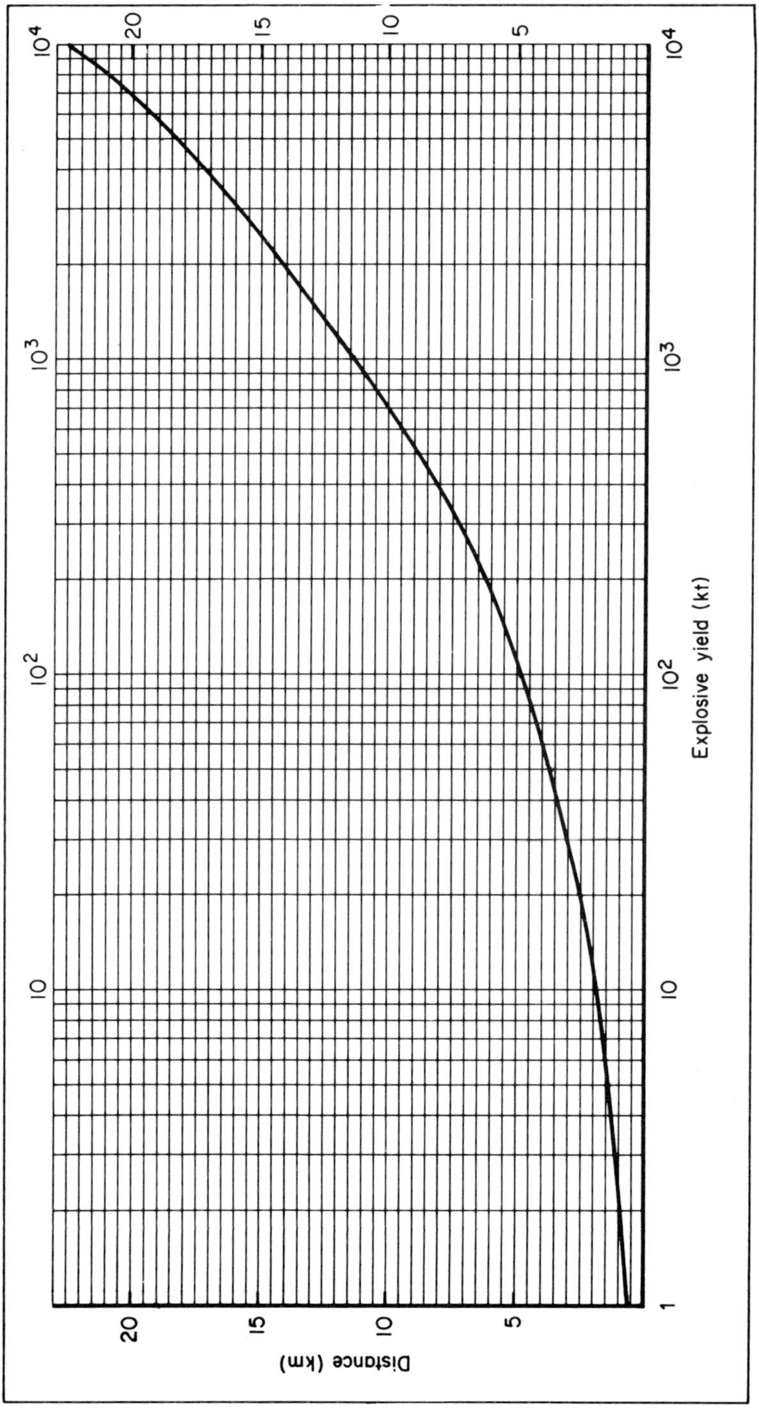

Figure 3. Slant distance from burst for third-degree burns

Apart from causing burns on exposed skin, the thermal radiation can also bring death and injury through fires. Flammable and combustible materials may be ignited by the heat pulse; these materials include furniture and fabrics inside buildings, or dry grass and leaves outside. The blast wave also contributes to starting fires: damaged furnaces and stoves, overturned electric radiators, broken gas pipes, and petrol in cars are all likely to start many fires. Under certain circumstances the individual fires may consolidate into one huge fire covering a large area; this fire may be a firestorm or a conflagration. In the firestorm the merging of many small fires causes a single convective column which blows inwards from an outside perimeter and creates temperatures of over 1 000 kelvin, so that everything combustible within the area is destroyed. In a conflagration, the fire front moves outwards and spreads as long as there is material to fuel it. The area and extent of damage due to these fires may far exceed those from the direct burns.

2.2.3. Blast wave

About half of the energy released in tropospheric explosions (i.e., at an altitude of less than about 15 000 metres) is in the form of the blast wave. If the bomb is exploded over a centre of population, blast will give rise to the largest number of immediate casualties; it is also the main cause of structural damage.

At the instant of the explosion the gases in the small region originally occupied by the bomb itself build up tremendous pressures, of the order of millions of times the atmospheric pressure. Owing to this pressure thc gases rapidly expand, giving rise to a pressure wave called a shock wave; if formed in air, it is referred to as a blast wave. The front of this wave spreads in all directions, with a speed several times greater than the speed of sound at sea level. As the blast wave progresses, its amplitude and velocity decrease. The actual sequence of events that follow the initial shock wave is highly complex, and a detailed description of it is outside the scope of this survey. Among the important manifestations are the formation of a crater, and the effects of overpressures in the blast wave.

If the fire-ball touches the ground, a crater is formed in which everything is vaporized. The size and depth of the crater depend on the yield of the bomb, on the height above ground of the burst, and on the nature of the soil.

Appreciable crater formation will occur when the height of the burst is less than about one-tenth of the maximum radius of the fire-ball. For a surface explosion the diameter of the crater in dry soil, or dry soft rock, is 45 metres for a 1-kt yield, and it increases by a factor of 2 for every tenfold increase in yield. For a bomb of a given yield, larger craters are produced with underground explosions, the largest crater occurring at a certain optimum depth.

At much greater distances from the explosion, impairment and death result from direct and indirect injuries produced by the blast wave. The human body can withstand quite high overpressures, over 200 kPa (2 atmospheres); most blast deaths occur from indirect effects—from the collapse of buildings, from being blown into objects or from falling debris. It is generally assumed

Figure 4. Distance from ground zero for 35 kPa overpressure

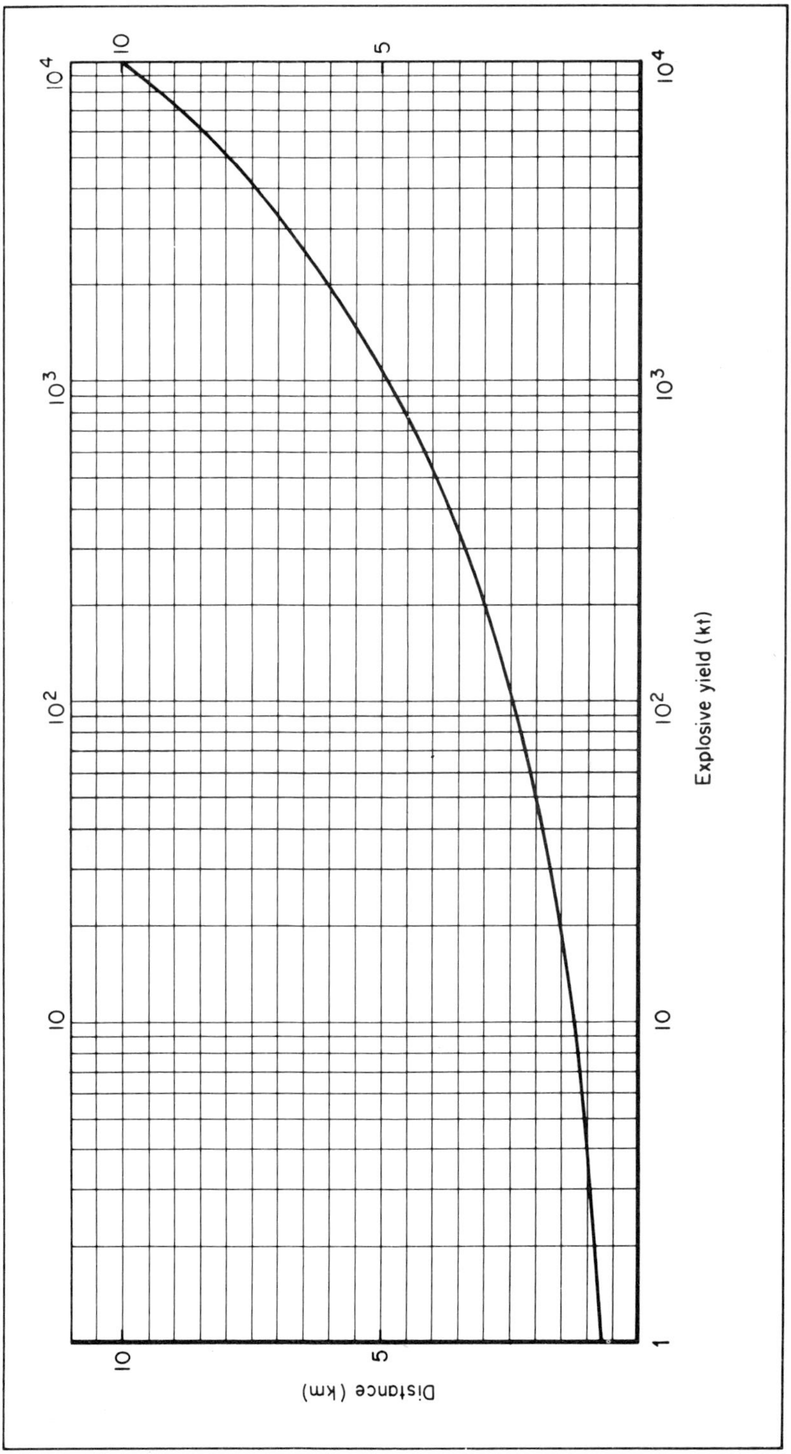

that an overpressure of 35 kPa is the relevant value for calculating the lethal effects of the blast wave. The area in which this overpressure occurs is called the lethal area for blast effects, and is defined as the circular region within which the number of survivors is equal to the number of persons killed by the blast outside this region. The total number of deaths due to blast effect can thus be calculated by multiplying this area by the population density (provided the latter is constant).

For a given explosive yield of the bomb there is an optimum height of burst for which a given overpressure will extend farthest from ground zero. For example, an overpressure of 35 kPa will be reached farthest from the point of explosion of a 1-kt bomb if the detonation occurred at a height of about 320 metres; the distance from ground zero for this overpressure is then about 700 metres. For other explosive yields the optimum height increases as the cube root of the yield; thus, for a 1-Mt bomb the optimum height for the burst will be about 3 200 metres.

For explosions at altitudes lower than the optimal height, the distance at which a given overpressure occurs is reduced, but much greater pressures are then produced at shorter distances from ground zero. If the objective of the bombing is the destruction of a hard target, such as a missile in an underground silo, the lower the altitude of the explosion the more effective it will be; surface bursts are therefore likely to be employed. As already mentioned, low-altitude bursts also give rise to intense early fall-out.

Figure 4 shows the distance at which a peak overpressure of 35 kPa occurs, for bombs exploded at a height of 300 metres above ground, as a function of the explosive yield of the weapon. Since a constant height was assumed, the distance for the 35 kPa overpressure does not increase with explosive yield as fast as it would for optimal heights (i.e., as the cube root of the explosive yield). Thus, for a 10-kt bomb the distance is about 1.3 km (lethal area about 5 km^2) whereas for a 10-Mt bomb the distance is about 10 km (lethal area over 300 km^2).

2.2.4. Electromagnetic pulse (EMP)

The initial ionizing radiation from a nuclear weapon causes intense ionization of the air; one of its consequences is a short but powerful pulse of electromagnetic radiation. The pulse is similar in character to that of radio waves, but covers a continuous spectrum of wavelengths, and with an amplitude millions of times stronger than that in ordinary radio waves. The pulse is over in a few microseconds, but despite its short duration it can interfere with and cause surges and permanent damage to electrical equipment within a distance of up to 20 km for a low-altitude explosion. For high-altitude bursts the area affected is greatly increased; it may reach out to distances of thousands of kilometres for high-yield explosions at a height of a few hundred kilometres. Electronic equipment based on solid state devices such as transistors is particularly sensitive to the effect of the EMP.

The EMP is not hazardous to human beings except for possible severe electric shocks to a few persons. It is indirectly relevant to human casualties, however, because of the likely disruption of electric power supplies, and telephone and radio communication, should equipment at the transmitting and/or receiving ends be burnt out. This may seriously affect the chances of survival of injured persons. Some types of radiation-monitoring equipment may also be put out of action or give false readings, thus adding to the problems in dealing with radiation exposure.

2.3. The nuclear arms race

2.3.1. Stockpiles of weapons and means of delivery

The two atom bombs dropped in 1945 on Japanese cities—the only ones to be used so far in combat—were fission bombs of low efficiency: the Hiroshima bomb, in which uranium-235 was the nuclear explosive, was detonated by the gun technique and had an explosive yield of 12.5 kt; the Nagasaki bomb, based on plutonium-239 and detonated by implosion, had nearly twice the explosive power, or a yield of 22 kt. Both were delivered by aircraft and were detonated at heights of 570 and 500 metres respectively [2]. By present-day standards, these bombs, which destroyed two cities and caused perhaps a quarter of a million deaths, appear archaic and diminutive. The advances made in weaponry during the past 36 years have been both qualitative and quantitative. The efficiency of fission weapons has been improved by more than an order of magnitude and, by the development of the F-F-F bomb, the total explosive and fission yields have been increased one thousandfold. The design of bombs has been specialized to suit specific requirements for strategic or tactical purposes, to enhance radiation lethality and minimize material damage, or vice versa. Above all, both great powers have built up their stockpiles of nuclear weapons to an incredible extent. Although the exact size of the stockpiles is not known, a minimum value can be deduced from the declared or estimated sizes of the various types of delivery vehicle and their warheads. Altogether, the explosive power in the strategic arsenals of the two great powers is estimated to be over 13 000 Mt (see table 3).

In parallel with the advances in the technology of the weapons themselves there has been colossal progress in developing the means of their delivery, making them more reliable and faster, and increasing their range and the precision of targeting. The biggest advance has been in the use of ballistic missiles which can be launched from underground silos or from submarines. Travelling at a speed of some 30 000 km per hour, these missiles have an intercontinental range of up to 13 000 km and can hit a target over this range after a transit time of about 30 minutes. Despite this huge distance, the aiming precision is so good that some of these missiles have a 50 per cent probability of landing within 200 metres of a stationary target.

The aiming accuracy is expressed by the term 'circular error probable' (CEP); it is defined as the radius of the circle, with the target at its centre, within which the missile has a 50 per cent probability of landing. The earlier

Table 3. US and Soviet strategic nuclear forces (estimated for 1985)

(a) USA

Launcher	Number	Warheads per launcher	Total warheads	Yield (Mt)	Total megatons	Equivalent megatons
ICBMs						
Minuteman II	450	1	450	1.0	450.0	450
Minuteman III	550	3	1 650	0.17	280.5	512
Titan II	54	1	54	9.0	486.0	232
Total ICBMs	1 054		2 154		1 216.5	1 194
SLBMs						
Poseidon	336	10	3 360	0.04	134	403
Poseidon C-4	160	8	1 280	0.10	128	282
Trident I	240	8	1 920	0.10	192	422
Total SLBMs	736		6 560		454	1 107
Bombers						
B-52 G/H	165	6 SRAM	990	0.2	198	337
		4 bombs	660	1.0	660	660
B-52 CM	165	20 ALCM	3 300	0.2	660	1 122
FB-111	60	2 SRAM[a]	120	0.2	24	41
		2 bombs	120	1.0	120	120
Total bombers	390		5 190		1 662	2 280
Totals	**2 180**		**13 904**		**3 332.5**	**4 581**

(b) USSR

Launcher	Number	Warheads per launcher	Total warheads	Yield (Mt)	Total megatons	Equivalent megatons
ICBMs						
S-11	330	1	330	1.5	495	432
SS-17	200	4	800	0.6	480	560
SS-18	308	8	2 464	1.5	3 696	3 228
SS-19	500	6	3 000	0.8	2 400	2 580
SS-16[b]	60	1	60	1.0	60	60
Total ICBMs	1 398		6 654		7 131	6 860
SLBMs						
SS-N-6, SS-N-8	600	1	600	1.0	600	600
SS-N-17, SS-N-18	300	3	900	0.2	180	306
Total SLBMs	900		1 500		780	906
Bombers						
Bear	100	1	100	20	2 000	740
Bison	40	1	40	5	200	116
Total bombers	140		140		2 200	856
Totals	**2 438**		**8 294**		**10 111**	**8 622**

[a] Short-range attack missile.
[b] The production of these has since been cancelled.

Source: Counterforce Issues for the U.S. Strategic Nuclear Forces (Congressional Budget Office, Washington, D.C., 1978).

missiles had a CEP of several kilometres. Advances in guidance systems have greatly reduced the CEP and with further improvements in terminal guidance, CEP values of 50 metres or less are envisaged.

Another important step in the nuclear arms race was made when missiles began to be equipped to carry a number of warheads, each capable of being separately fired and directed onto a different target while in flight. These are the so-called multiple independently targetable re-entry vehicles (MIRVs); an increasing proportion of the nuclear strategic forces is being put into this category.

MIRVing of missiles has made it possible to increase the number of warheads by an order of magnitude, for the same number of launchers, and has brought the total number of warheads in the strategic forces of the great powers to over 22 000.

This can be seen in table 3, published in 1978 by the US Congressional Budget Office [3a], which shows the strategic nuclear forces of (*a*) the United States and (*b*) the Soviet Union as estimated to be deployed in 1985. Both great powers deploy three systems of weapons, launched from land, sea and air. The systems include: intercontinental ballistic missiles (ICBMs) launched from underground silos, submarine-launched ballistic missiles (SLBMs), and missiles launched from aircraft. The characteristics of some of these missiles are given in table 4.

Table 4. Characteristics of some nuclear missiles

Missile	Number of re-entry vehicles	Explosive yield per re-entry vehicle (kt)	Range (km)	CEP (m)
US missiles				
Minuteman III	3	170	13 000	≲300
Poseidon	10	40	4 600	~500
Trident I	8	100	7 500	~500
Soviet missiles				
SS 17	4	500	9 000	~500
SS 18	10	1 500	10 000	~300
SS 19	6	500	9 000	~300
SS-N-8	1	1 000	7 800	1 300
SS-N-18	3	200	9 500	~500

Source: SIPRI.

As is seen from table 3, many of the ICBMs and SLBMs are MIRVed with up to 10 warheads per launcher. Some of the US long-range bombers carry 20 air-launched cruise missiles (ALCMs). As a result of the greatly improved aiming accuracy, a lower explosive yield is needed to destroy a target; for this reason the MIRVed warheads have low yields, down to 40 kt. But both sides still have bombs of high yields: the 9-Mt bombs carried by the

US Titan ICBMs, and the 20-Mt bombs carried by Soviet Bear bombers. The aiming accuracy is still better, on the whole, in the US missiles, and largely because of this the average yield of these missiles is considerably lower than that of the Soviet warheads. The total equivalent megatonnage of the latter's strategic forces is thus nearly twice that of the United States, despite the smaller number of warheads. As mentioned earlier, from the point of view of the destruction caused by the blast effect the equivalent megatonnage (shown in the last column) is the important parameter. However, because of the mixture of low and high yields, the total equivalent megatonnage held by both sides comes out to be nearly the same as the total actual megatonnage.

In addition to the strategic forces, both nations also have huge arsenals of so-called tactical nuclear weapons although, in an almost continuous spectrum of types, yields and ranges, the distinction between tactical and strategic weapons is not always clear. Tactical weapons include shells which can be fired from 155-mm howitzers and other artillery. The yields vary from 0.5 kt upwards, but weapons with even lower yields, between 10 and 100 tons of TNT, are in existence. These are the so-called 'mininukes'. The ranges of tactical weapons vary from 3 to 160 km. The United States is thought to possess a total of 22 000 tactical nuclear weapons and the Soviet Union about 8 000.

In between strategic and tactical weapons are the so-called theatre nuclear forces, or grey area weapons. To this category belong the US cruise missiles, pilotless planes which can be launched from aircraft (ALCMs), from the ground (GLCMs), or from submarines (SLCMs), and the Soviet Union's Tu-22M (Backfire) bomber aeroplanes. The cruise missiles can have a range of up to 3 000 km and a CEP of less than 30 metres and can carry up to 4 nuclear warheads of 100 kt each. The Backfire bomber has a range of 7 500 km and can travel at a speed of Mach 3 with a 9-ton payload.

The other three nuclear powers, China, France and the UK, have both fission and thermonuclear weapons in their arsenals. These arsenals are very much smaller than the stockpiles of the two great powers, but little is known about the nuclear potential of China. Unlike the UK, a member of NATO, France's formal stand is as an independent nuclear power although in any war France is likely to be in alliance with the United States. China, however, has definitely emerged as a third contender in the nuclear arms race, particularly after demonstrating the possession of intercontinental ballistic missiles.

The SALT (strategic arms limitation) talks have been conducted by the USA and the USSR as if no other nuclear powers existed, but some NATO countries are becoming increasingly worried about the balance of power in Europe and want to have a say in further negotiations. The European situation has become aggravated by the NATO decision to modernize its theatre nuclear forces by the deployment in Europe of 108 Pershing II missiles and 464 ground-launched cruise missiles; this has brought strong protests from the Warsaw Treaty Organization and may lead to a perilous situation. The official reason for NATO's modernization programme is to provide a balance to the deployment by the USSR, in increasing numbers, of the SS-20, a mobile

MIRVed ballistic missile with a range of up to 4 000 km. But the Soviet Union claims that the SS-20 is not a new weapon but only a replacement of its old-fashioned SS-4 and SS-5 missiles. This is an example of the action-reaction syndrome which exacerbates the nuclear arms race and keeps it accelerating.

2.3.2. Vertical proliferation

The total size of the nuclear arsenal is uncertain, but table 5, from a recent UN Report [4a], gives a rough estimate of the total number of warheads and their total explosive yield. The numbers are rounded to the nearest thousand for the arsenals of the two great powers and to the nearest hundred for the other nuclear states. The ranges of values reflect the uncertainty of the estimates.

Table 5. Current nuclear arsenals

Nation	Strategic	Other systems	Total
(a) Number of warheads			
USA	9 000–11 000	16 000–22 000	25 000–33 000
USSR	6 000– 7 500	5 000– 8 000	11 000–15 500
UK			200– 1 000
France			<200
China			<300
Total			**37 000–50 000**
(b) Explosive yield (Mt)			
USA	3 000– 4 000	1 000– 4 000	4 000– 8 000
USSR	5 000– 8 000	2 000– 3 000	7 000–11 000
UK			200– 1 000
France			200– 400
China			<100
Total			**11 000–20 000**

Source: Comprehensive Study of Nuclear Weapons, United Nations Report A/35/392, Report of the Secretary General (United Nations, New York, 1980), p. 31.

Estimates arrived at on a different basis lead to even higher numbers, but even the numbers in the table show that the total megatonnage amounts to 5 tons of high explosive for each inhabitant of the Earth, an unprecedented and awesome accumulation of destructive potential.

Even so, a new escalation of the arms race is in progress. The MIRVing of missiles and the great accuracy in aiming have made land-based missiles in fixed silos highly vulnerable. They evoked the fear of a first-strike attack in which the elimination of land-based missiles and bombers would reduce the retaliatory potential to such an extent that the deterrent would no longer be effective. To overcome this the USA is to develop a new system of missiles, the MX (missile experimental). Apart from greatly improved overall performance, the main feature of the MX system is mobility. Twenty-three locations would be provided for each of the MX missiles; each missile could be moved from one location to another by means of a huge transporter-vehicle. In this way it is hoped that the probability of a missile being hit would be reduced to 1 in 23. In the light of the past history of the nuclear arms race it seems likely that the Soviet response to the MX system, apart from developing a similar system, would be to increase the number of warheads twentyfold.

The vulnerability of land-based missiles has brought about another danger, the launch-on-warning response. As soon as a signal is received that an enemy's missiles have been launched and are on the way, all land-based missiles will be launched in a counter-attack, since they would otherwise be destroyed. This strategy introduces a further element of instability, the danger of a war starting by accident due to a false alarm.

The continuous qualitative and quantitative progress in nuclear weapons, far from bringing greater security, has in fact resulted in a destabilization of the situation; there is less security and an increased danger of a nuclear war.

2.3.3. Horizontal proliferation

A further increase in the probability of a nuclear war is bound to result from horizontal proliferation, that is, the acquisition of nuclear weapons or a nuclear weapon potential by countries which are not designated as nuclear weapon states. India was the country to start this new category by testing a nuclear device in April 1974. Ostensibly this test was for the purpose of exploring the possibility of peaceful nuclear explosions, but no further efforts appear to have been made to justify this excuse. Soon after India's test, pressure began to build up in Pakistan for the acquisition of nuclear weapons; this has not yet been achieved, but Pakistan is making strenuous efforts to acquire facilities, such as enrichment and reprocessing plants, for making nuclear weapon materials, allegedly for its peaceful nuclear power programme. This example illustrates how real the danger of the spread of nuclear weapons is: India no doubt felt that possession of the bomb would strengthen its position *vis-à-vis* a neighbouring nuclear power, China; and Pakistan reacted in the same way in relation to India.

It is generally believed that Israel has built its own nuclear weapons; possibly related to this are the attempts by Libya and Iraq to acquire such weapons, with the bizarre claim that there is a need for an 'Islamic' bomb. South Africa is also believed to have developed a nuclear weapon capability; its intention to carry out a nuclear test in the Kalahari Desert may have been abandoned after a strong representation from the United States, but there are rumours that South Africa carried out a clandestine test in the Indian Ocean. Other potential nuclear weapon countries include Argentina, Brazil and Cuba, which are engaged in 'peaceful' nuclear activities but which have not signed the Non-Proliferation Treaty (NPT).

Even if all these countries and their neighbours were to acquire nuclear weapons in the foreseeable future, the addition to the nuclear arsenals already in existence would be insignificant in terms of numbers, but the peril to world security would increase considerably. The fact, or even rumour, of more countries 'going nuclear' would certainly create greater tension and could itself be a cause of conflict. More importantly, in a world divided between two (or among three) ideologies, there is always the possibility of a small country appealing to one of the great nuclear powers for help; the latter may be obliged by treaty or other bond to come to the assistance of the smaller country. Thus, a nuclear war begun on a very small scale, in a remote part of the world, carries the risk of escalating into a nuclear war between the great powers.

2.4. *War scenarios*

A nuclear war could occur in an infinite variety of ways, from the field use of nuclear artillery to stop the advance of tanks, to an indiscriminate blitz with nuclear strategic warheads on the whole territory of the adversary. In attempts to estimate the consequences of a nuclear war a variety of scenarios have been enacted and many kinds of war games played. In these games different postures are assumed, depending on the particular type of strategic policy which happens to be in vogue at a given time. The so-called counterforce strategy, now the official policy of the United States, envisages an attempt by one side to knock out the fighting potential of the other side; this means aiming at the destruction of all known missile bases but not at centres of population. Even such a limited objective would result in a vast number of civilian casualties by radiation from fall-out. But in any case, an enemy wounded but not knocked out completely is likely to retaliate with bombs not on missile bases—which would by then be empty—but on centres of population. The whole concept of a limited nuclear war is a dangerous fallacy; one side can start a nuclear war, but it takes two sides to keep it limited. It is delusory to think that a nuclear war would be played according to some sort of Queensberry Rules and that it would be kept limited in scope, for example to the use of tactical or theatre nuclear weapons only. Just as a war started with conventional weapons is likely to escalate into a nuclear war—particularly if many nations possess nuclear weapons—so a tactical nuclear war has a high probability of escalating into a wholesale strategic nuclear war. On a realistic basis it must be assumed that the worst would happen, once a nuclear war had started.

3. Biological effects of radiations on man

It is generally accepted that any exposure to ionizing radiation may produce harmful effects, but the type and the severity of the effect and the time of its appearance vary considerably, depending primarily on the dose of radiation. With high doses the symptoms of exposure, which may lead to death, are noticed shortly after the exposure. These are known as acute effects, and have to be distinguished from the long-term effects which may follow exposures to medium-high and to low doses, and which themselves fall into two categories: somatic and genetic. Somatic effects are those experienced by the individual exposed; some of these may not become apparent for many years after the exposure and they may take different forms, the most prominent being the induction of cancer. Genetic effects are those occurring in the children of the individual exposed or in future descendants, and may appear in a variety of defects which are described as hereditary.

The biological effects of radiation have been the subject of intense study for many years and it has been claimed that more is known about radiation hazards than about all other environmental or occupational hazards of modern society. Yet when it comes to the quantitative estimate of the harm to man from a given dose of radiation there are very large uncertainties, sometimes amounting to an order of magnitude; this applies to both acute and long-term effects. A recent comprehensive study of the harmful effects of low doses of radiation, by a committee of the US National Academy of Sciences, began its report with the following words of caution: "the risk estimates presented here should in no way be interpreted as precise numerical expectations. They are based on incomplete data and involve a large degree of uncertainty, especially in the low-dose region" [5]. It should be added that in the high-dose region too there is a large degree of uncertainty, mainly due to the absence of sufficient data.

Although it is impossible to say whether any new information, as it becomes available, will prove present estimates of hazards to have erred on the high or on the low side, past experience indicates that the more we learn about the effects of radiation the greater our concern becomes. The general trend has been towards lowering the value of the dose which is thought to produce a given level of risk. This is reflected in the regulations governing civilian radiation protection for occupational and population exposures. The maximum permissible doses, or the 'dose limits' in current terminology, have fallen dramatically in the past, and although they have remained unchanged in recent years this is because previously built-in safety margins have been removed;

effectively this means a further drop in dose limits.

For these reasons calculations of radiation casualties in warfare are more likely to underestimate than to overestimate the hazards.

3.1. Radiation doses

3.1.1. The units of radiation dose

The dose of radiation is measured in terms of the energy of the ionizing radiation absorbed in tissue. The unit of dose is the gray (Gy) and it corresponds to the absorption of 1 joule per kilogram of tissue (1 Gy = 1 J kg^{-1}). The former unit of dose still used in many publications was the rad; it is equal to one hundredth of a gray (1 Gy = 100 rad).

The radiation dose must be distinguished from another quantity applicable to X-rays and gamma-rays—radiation exposure. It is a measure of the ionizing action of the radiation on a standard substance, air. (Air was originally chosen because ionization in it was easy to measure.) Radiation exposure is expressed in terms of the electric charge produced by the radiation in a given mass of air. The unit of exposure is the quantity of radiation which produces ions of a total charge of 1 coulomb in 1 kilogram of air. The former unit of exposure was the roentgen (R). Many of the instruments used to monitor the dose of radiation are based on the measurement of ionization in air and are often calibrated in roentgens.

The relationship between the rad and the roentgen is not simple; it depends on the physical properties of both the radiation and the medium through which it passes. For the gamma-rays likely to occur in nuclear warfare and for doses measured in air, 1 R = 0.87 rad. For water and soft tissue the rad and the roentgen are even closer, namely 1 R = 0.96 rad. However, although they have nearly the same numerical value, it should be remembered that the rad (or the gray) and the roentgen are units of different quantities.

Even for the same physical dose, the biological effect may be different for different types of radiation. For our purpose radiations can be roughly divided into two groups, low and high linear energy transfer (LET). LET refers to the ionizing powers of the radiation. Of the radiations encountered in nuclear warfare, beta-rays, gamma-rays, and X-rays are classified as low-LET, while neutrons and alpha-particles are high-LET radiations. For the same dose, the high-LET radiation generally produces more biological damage than low-LET radiation. The ratio of the doses of low- and high-LET radiation, which produce the same degree of biological damage, is called the relative biological effectiveness (RBE) (see section 3.6.1) or the quality factor.

The dose of low-LET radiation which is estimated to be equivalent in biological effect to a given dose of high-LET radiation is called the 'dose equivalent' of the latter and is measured in sieverts (Sv). One sievert of high-LET radiation, or any mixture of radiations, is thus assumed to be equivalent in its effect to one gray of low-LET radiation. The former unit of dose equivalent was the rem (1 Sv = 100 rem). Therefore, when calculating the effect caused by several types of radiation, for example gamma-rays and neutrons, the sievert (or the rem) is the appropriate unit.

The various old and new units and the relations between them are set out in table 6.

Table 6. Radiation units

Quantity	Present unit	Former unit	Relation between them
Dose	gray (Gy)	rad	1 rad = 0.01 Gy
Dose equivalent	sievert (Sv)	rem	1 rem = 0.01 Sv
Exposure	($C\ kg^{-1}$)	roentgen (R)	1 R = $2.58 \times 10^{-4} C\ kg^{-1}$
Activity	becquerel (Bq)	curie (Ci)	1 Ci = 3.7×10^{10} Bq

In addition to the dose itself, the rate at which it is delivered has an important bearing on the biological effects produced. In this connection one can speak of acute exposure when the time of the exposure to radiation is short, and chronic exposure when it is protracted over a long time. The dose rate is defined as the dose received per unit of time.

Another relevant quantity, particularly when many people are exposed to radiation, is the collective dose. It is the product of the dose and the number of people exposed to it; thus, it is a measure of the total burden on the population due to the exposure to radiation. The collective dose is measured in man-gray and the collective dose equivalent in man-sievert.

3.1.2. Radiation doses to organs in the body

A person may be exposed either externally, when the source of radiation is outside the body, or internally, if he inhales or ingests a radioactive substance. Most of the exposures encountered in nuclear warfare result from external irradiation, but internal exposures become predominant in some circumstances (see sections 3.3.7 and 3.7).

A division can also be made between whole-body (or total-body) exposure, when all organs are exposed (not necessarily uniformly), and partial-body exposure, when only one or a few organs are exposed. Internal sources usually give rise to partial-body exposure.

In the case of external exposure the body may be irradiated from many directions or from one direction only. An example of the former is when the exposure is due to fall-out deposited on the ground. Uni-directional exposure results from the initial radiation from the bomb. In both cases, but especially in that of uni-directional exposure, it is important to distinguish between the amount of radiation reaching the surface of the body and that reaching the various organs inside the body. These two quantities differ because of the physical interactions between the radiation and the atoms of the body as the radiation passes through it. In general this results in an attenuation of the intensity of the radiation. The degree of attenuation depends on the properties of the radiation (type, energy) as well as on the size of the body and the depth of the given organ.

When considering the injury caused by exposure to radiation it is the dose to the given organ or—in the case of whole-body exposure—the mean dose to all tissues (which might be presented by the midline tissue dose) that is the determining quantity. Nevertheless, the radiation doses which are usually quoted in descriptions of the effects of nuclear explosions refer to the tissue dose measured at the surface of the body.

The distinction between surface tissue doses and midline tissue doses, or organ doses, is important when the population exposed consists of individuals of different sizes, including children. For example, for the gamma-rays encountered in nuclear warfare the mean dose to the active bone marrow (a critical organ for the chances of survival after exposure) of an adult weighing 70 kg is 53 per cent of the surface tissue dose in uni-directional exposure; for a five-year old child weighing 20 kg the dose is 63 per cent; and for a newborn baby weighing 3.9 kg the internal dose is 70 per cent of the surface tissue dose [6]. For neutrons, which are much more attenuated in the body than gamma-rays, the differences are even larger. When combined with the greater intrinsic sensitivity of children to radiation (see section 3.6.4) this can literally mean the difference between life and death. A level of external radiation exposure which produces hardly any mortality in adults may give 100 per cent mortality in children. Altogether the chances of survival of babies and infants are very much smaller than for adults exposed at the same radiation level.

3.2. Natural sources of radiation

Before discussing the biological effects produced by different doses of radiation it is useful to be reminded that man has always been exposed to radiation, the so-called natural background. Radiation reaches the human body from different sources, the chief ones being: cosmic radiation, which comes from outer space; terrestrial radiation, from the radioactive substances contained in the Earth's crust, some of which are incorporated in building materials; and internal radiation, mainly from potassium-40, a radioactive element which is a natural constituent of the human body, but also from other radioactive elements inhaled or ingested. Table 7 shows the contribution made by these sources to the tissues in different organs of the body. The large internal dose to the lung is due to the inhalation of a radioactive gas, radon-222, a component of terrestrial radioactivity; radon and some other nuclides emit alpha-particles which have a high RBE, resulting in a somewhat higher dose equivalent locally in a few organs.

It is seen that the total natural background is about 1 milligray per year. This is an average value for people living near sea-level; with increasing altitude there is a slight increase in natural radiation, owing to the more intense cosmic

Table 7. Natural background of radiation (micrograys per year)

Source	Bone marrow	Bone lining cells	Lung	Gonads
External				
Cosmic rays	280	280	280	280
Terrestrial	320	320	320	320
Internal				
Potassium-40	270	150	170	150
Other nuclides	50	90	330	30
Totals	**920**	**840**	**1 100**	**780**

Source: Sources and Effects of Ionizing Radiation, United Nations Scientific Committee on the Effects of Atomic Radiation (United Nations, New York, 1977), p. 11.

radiation. In a few small regions of the world, where the radioactive content of the Earth's crust is large, the terrestrial component is enhanced, giving a natural background 10–20 times greater.

Additional to the natural background radiation are exposures from various man-made sources. Chief among them are the radiation doses from medical examinations or treatment; there is still also a significant contribution from the past testing of nuclear weapons in the atmosphere. In the developed countries the addition from these man-made sources amounts to some 30–70 per cent of the natural background.

3.3. Acute effects of radiation

The symptoms of acute effects appear soon after the exposure to radiation, within an hour or two (or even within minutes after very high doses), but death as a result of the exposure may not occur for some time. It is generally believed that death from acute effects comes within two months after exposure, but there is evidence from the bombings in Japan that it may be much later, up to several years. By that time deaths from long-term effects, in particular leukaemia, begin to occur, so there is no clear-cut division between acute and long-term effects in the time of occurrence of death.

Acute effects may manifest themselves when the whole body or a large part of it is exposed in a short time to a radiation dose from about 0.2 Gy upwards. The early symptoms, such as anorexia, nausea and headaches, are part of the so-called prodromal syndrome. With doses up to 1 Gy these symptoms soon disappear and recovery is apparently complete. With increasing dose mortality increases, reaching 100 per cent for a dose of about 5 Gy (midline tissue dose), although in healthy adults survival is possible even for somewhat larger doses if special treatment is provided. Death in the dose range 1–5 Gy is mainly due to damage to the blood-forming organs. Larger doses are invariably fatal, death being due to disturbances of the gastro-intestinal system and—at still higher doses—to damage to the central nervous system.

The time of death depends on the size of the dose. In figure 5 the time of death is plotted as a function of the radiation dose given to the whole body. It should be noted that the curve is largely based on data from experiments with mammals. Although there were many deaths due to acute radiation exposure in Hiroshima and Nagasaki, there is no record of the doses received. A few points on the curve were obtained from the small number of radiation deaths due to accidents, but the overall shape of the curve must be considered as only a gross approximation.

Figure 5. Time of occurrence of death from acute effects

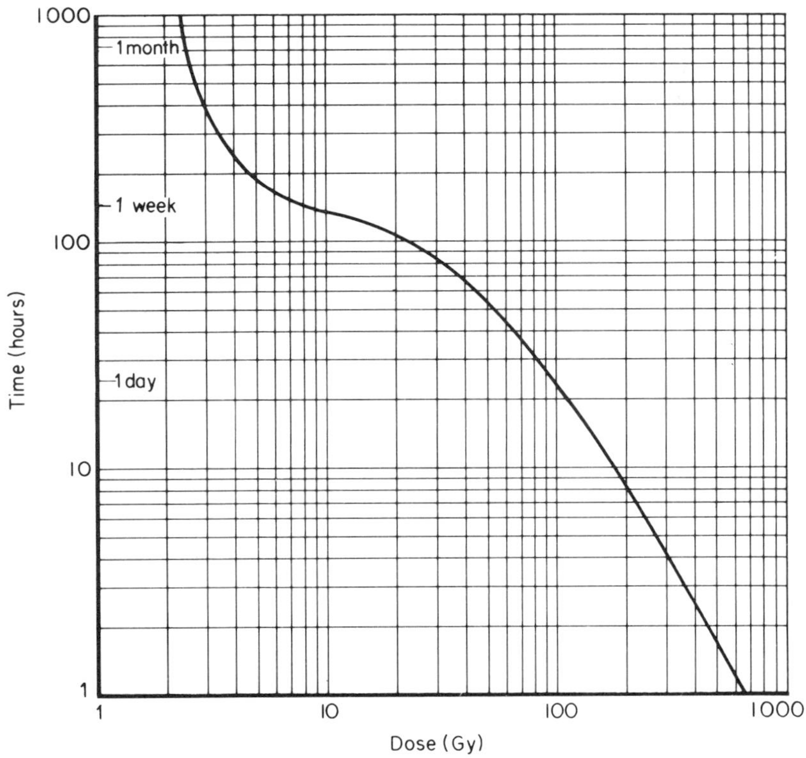

It is seen that the curve divides roughly into three components. At the high-dose range, from about 20 Gy upwards, the time of occurrence of death decreases rapidly with increasing dose; at very high doses death may occur in hours, or even in minutes. The second component covers doses between about 5 and 20 Gy; here the time of death varies much less with dose and is of the order of one week. In the third component, from 5 Gy downwards, the time of death again varies faster with the dose; it ranges from a few weeks to some months, with increasing probability of survival as the dose decreases.

The three components of the curve correspond to the three aforementioned specific forms of the acute radiation syndrome, namely haemopoietic, gastro-intestinal and cerebral.

3.3.1. The prodromal syndrome (radiation sickness)

Soon after exposure to radiation, a person may begin to show symptoms of acute gastro-intestinal and neuro-muscular effects. These are popularly known as radiation sickness but are more correctly described as the prodromal syndrome. The gastro-intestinal symptoms are: anorexia, nausea, vomiting, diarrhoea, intestinal cramps, salivation, dehydration and loss of weight. The neuro-muscular symptoms include fatigue, apathy or listlessness, sweating, fever, headache and hypotension followed by hypotensive shock [7]. All these symptoms occur only at high doses; with low doses, and during the first 48 hours after the exposure, only some of the symptoms may occur.

Many of the above symptoms may also be caused by factors other than exposure to radiation. The state of nervous tension when war breaks out, or crowding in shelters, may simulate radiation sickness. Witnessing genuine symptoms of radiation exposure in others may evoke similar symptoms, with epidemic results, particularly in young children. Without a physical measurement of the radiation dose received by a person it would be difficult to distinguish between genuine and spurious reactions.

But even if the dose of radiation were known, it would be impossible in many cases to predict the outcome of the exposure in any individual. Very little is known about the mechanism which underlies the occurrence of the prodromal syndrome, but there is evidence of a large variation among people in their response to an exposure. This applies to all radiation effects. It has been suggested that in a given population there exist sub-groups with a higher than average sensitivity to radiation. The reason for the different sensitivity is unknown; it may be due to the genetic make-up, to the general state of health, or to some other cause. In any case, when many people are exposed to the same dose of radiation—in the lower range of the prodromal syndrome—some will show symptoms of radiation sickness but others will not. Therefore, in order to calculate the probability of a given symptom occurring, one has to take a statistical approach in terms of the percentage of an exposed population which will exhibit the symptom.

Figure 6 gives an example of such an approach applied to the mildest expression of the prodromal syndrome, anorexia [8a]. The vertical scale gives the percentage of persons exposed who exhibit anorexia, as a function of the dose. The horizontal bars at the 10, 50 and 90 per cent levels are a measure of the statistical error (95 per cent confidence limits). As is seen, a dose in the range 0.2–0.6 Gy is likely to produce anorexia in 10 per cent of the exposed population, a dose range of 0.7–1.3 Gy in half of the population, and a dose range of 1.7–4.4 Gy in 90 per cent of the population.

Table 8 gives the mid-line tissue doses for the 10, 50 and 90 per cent occurrences of other prodromal symptoms. Thus, a dose of 0.5 Gy is likely to

Figure 6. Probability of occurrence of anorexia

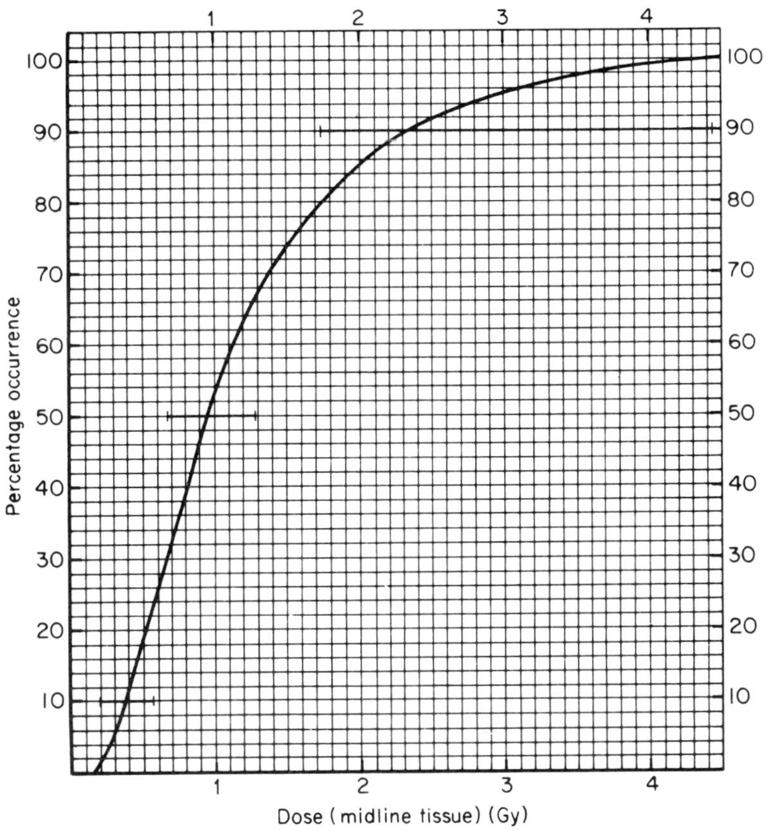

induce nausea in 10 per cent of persons exposed; a dose of 2.1 Gy will cause vomiting in half of them; and a dose of 3.9 Gy will cause diarrhoea in 90 per cent of those exposed as well as all the other symptoms in more than 90 per cent.

Table 8. Radiation doses (in grays) which produce early radiation sickness symptoms

Symptom	Percentage of exposed population 10%	50%	90%
Anorexia	0.4	1.0	2.4
Nausea	0.5	1.7	3.2
Vomiting	0.6	2.1	3.8
Diarrhoea	0.9	2.4	3.9

Source: Radiobiological Factors in Manned Space Flight, ed. W.H. Langham (National Academy of Sciences, Washington, D.C., 1967), p. 248.

This sequence of the symptoms, and the time of their onset and severity, may provide some guidance to the dose (if unknown) and to the prognosis. The higher the dose the earlier the onset and the more severe the symptoms. At moderate doses the symptoms subside after a few days and the person exposed may then feel quite well for a week or so. But subsequently they reappear and together with other signs of malaise may lead to death if the dose was high enough. Protracted vomiting during the first two days, and in particular the occurrence of diarrhoea, indicate a bad prognosis. Persons with intractable nausea, vomiting and diarrhoea will most likely die.

3.3.2. Effects on the blood-forming system

The main lethal effect of radiation is produced by the damaging action on the cells of which the body is made. With high doses the energy imparted to the cells is sufficient to destroy them completely, but much lower doses can inhibit mitosis, the ability of cells to divide. For the proper function of the organism it is vital that certain types of cell, which are continually lost, should be replenished by mitosis. But if after exposure too few dividing cells are left, the organism will not be able to function and death will occur. The most important of the cellular effects of radiation occur in the blood-forming tissues, particularly in the bone marrow.

Bone marrow stem cells and lymphocytes belong to the most sensitive components of the human body, as far as radiation is concerned. For example, the sensitivity of lymphocytes is so high that even a dose of 0.1 Gy will lead to structural abnormalities in them, and a dose of 2.5 Gy will reduce their number to less than 10 per cent. The other circulating blood cells are less radiosensitive, but they have to be continually renewed, and the destruction of bone marrow cells leads to an inhibition of the formation of all types of blood cell. Perhaps the most striking of the changes associated with radiation exposure is the reduction in the number of white blood cells. The purpose of these cells is to resist infection and remove toxic products from the body; a drastic reduction in their number will make the organism vulnerable to any infection. The platelets, which play an important role in clotting of the blood, show an even greater decrease in number for a given dose, leading to haemorrhage and purpura. Reduced production of red blood cells leads to severe anaemia.

In the dose range 1–5 Gy, the prodromal syndrome is followed by other clinical symptoms if too many bone marrow stem cells have been destroyed. These symptoms are haemorrhage under the skin, bleeding in the mouth and bleeding into internal organs. There is greater susceptibility to infection which causes a step-wise rise in temperature; severe emaciation and delirium lead to death, usually within six weeks.

3.3.3. The median lethal dose (LD-50)

As in the case of the prodromal syndrome, the response to exposure to a given dose within the range 1–5 Gy differs from individual to individual. Some will die but others will survive, and after some months the number of blood cells will be back to normal. The cells will show an apparent recovery although they will still be at risk from long-term effects. In many cases recovery will take place after a long time and after possible occurrence of such diseases as pancreatitis and liver function changes.

To calculate the probability of death (or survival) the statistical approach is again necessary. In this case, however, the amount of factual evidence available is much smaller. Figure 7 shows the percentage of persons, out of a large number exposed to a given dose of radiation, who are likely to die as a result of such an exposure [8a].

Figure 7. Probability of death from acute effects

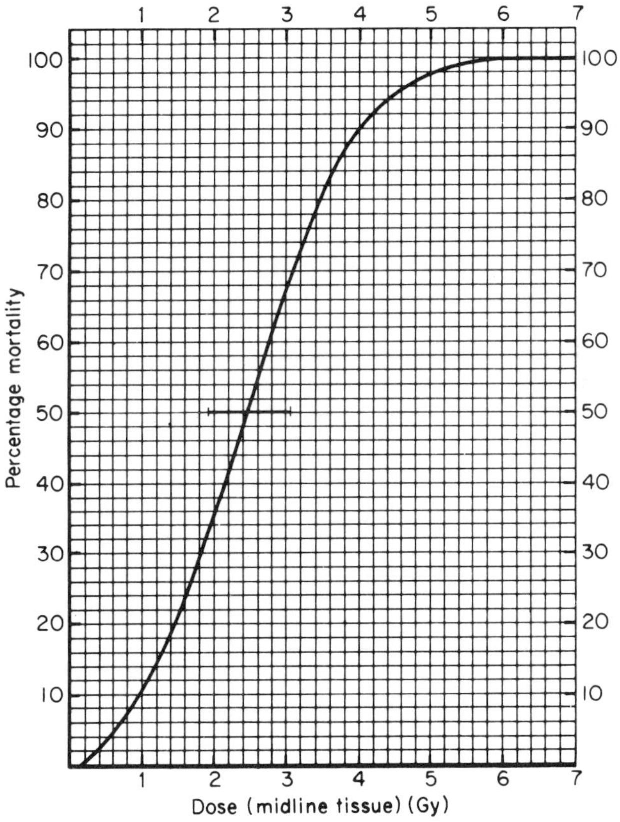

The mid point of the curve—the dose which gives a 50 per cent probability of death (or of survival) within a few weeks after exposure—is called the LD-50 dose (median lethal dose). It is an important parameter. LD-50 values have been measured carefully for many mammals and other living organisms, but for man it has been deduced from a small number of observations. The value in figure 7, about 2.5 Gy, may therefore be subject to a larger error than indicated by the bar at the 50 per cent level, which represents only the statistical error in fitting the curve.

A remarkable feature of the curve is its steepness. Within a very narrow range of doses, from 0.9 to 4 Gy, the probability of death increases from 10 per cent to 90 per cent. The error in measuring the dose (±50 per cent) nearly covers this range, so that it would be difficult to say who received a lethal dose. Combined with the uncertainties in the LD-50 value and in the slope of the curve, this makes any estimate of the number of survivors after an exposure to a dose from 1 to 5 Gy very dubious.

It should be noted that figure 7 applies to whole-body exposure of adults who—although not receiving special treatment—would be under care, particularly to avoid infection after the exposure. In conditions of nuclear war this is unlikely to be the case. Many of the persons exposed would also suffer from burns or mechanical injuries; these circumstances make infection more probable. In such situations the whole curve on figure 7 would shift to the left. The extent of the shift is impossible to predict but any exposure above 1 Gy might result in death under these circumstances.

3.3.4. Effects on the gastro-intestinal system

In the range of doses 5–20 Gy, death results earlier from changes in the gastro-intestinal system. Within minutes of the exposure there is a decrease in the rate of cell division in the lining of intestinal vessels; later the cells degenerate and a large proportion of them die. Without replacement of the cells the villi shrink; ulceration and gangrenous inflammation follow. There is high fever and persistent diarrhoea, rapidly progressing from loose to bloody stools. Abdominal distention and dehydration follow and the person becomes comatose. Death results principally from enteritis following infection, toxaemia and disturbance of body fluids. The chances of recovery after doses in this range are nil, except near the low limit of the range if special treatment is provided.

3.3.5. Effects on the central nervous system

At very high doses of 20 Gy and upwards, prompt death results from pathological changes in the nervous system. These include degeneration of brain cells, cerebral oedema and inflammation of cerebral vessels. After the initial phase of radiation sickness, as described above, there is a rapid progression from drowsiness to severe apathy and lethargy. There is generalized muscle tremor, lack of muscular co-ordination, coma, convulsions and shock. Persons exposed in this range of doses would be disoriented and incapacitated immediately, even if death were not to occur for a few days. Neutron doses in this range are planned to be used in the neutron bomb as a military weapon against tank crews (see section 4.1.4).

3.3.6. Factors affecting survival after acute exposures

The various factors which modify the biological response to radiation are discussed in section 3.6; this section describes briefly the factors modifying the acute effects, in particular the LD-50 value.

Like the absolute value of the LD-50, the influence of the physical and biological factors on this value is not known sufficiently to be expressed quantitatively with any accuracy.

The main physical factors are radiation quality and dose rate. It is generally assumed that at high doses the effectiveness of neutrons is about the same as of gamma-rays for the same dose in rads, but there is evidence that the RBE of neutrons is greater than unity even at high doses [8b]. However, a higher RBE would to a certain extent compensate for the greater attenuation of neutrons in passing through the body so that, for the same surface tissue dose, the acute effects of neutrons can be assumed to be roughly the same as of gamma-rays.

On the other hand, dose rate does affect markedly both the occurrence of the prodromal symptoms and the LD-50 value. From radiotherapy experience it is known that protraction of the exposure, either by giving it at a low rate or by dividing the dose into a number of fractions separated in time, requires an increase in the total dose in order to produce the same effect as a dose given in a short time. This indicates that the radiation damage is to some extent repaired in the time between the fractions or during the protracted exposure. In the case of continuous exposure the observations show that if the total dose was given in eight days instead of in one day all the values in table 8 for the prodromal symptoms would have to be doubled.

With regard to the LD-50 value, the curve on figure 7 can be applied if the total dose is delivered within one day or less. If the same dose is spread out over 2 weeks (the critical period for exposures from local fall-out) the LD-50 value will be nearly doubled. A further increase by a factor of 2 should be applied to the LD-50 value if exposure is protracted over a period of one year [8c].

Of the biological factors, the most important is the proportion of the volume of the body exposed to the radiation. Parts of the body, for example limbs, can be exposed to much higher doses before the acute symptoms occur. In total-body exposure an increase of the LD-50 value would also ensue if part of the body were shielded from the radiation or received a smaller dose. The greater the proportion of bone marrow screened from radiation, the greater the chance of survival.

Age at exposure is another relevant factor. The sensitivity to radiation appears to be greater both in the very young and the very old; for these population groups the LD-50 value is likely to be smaller than for adults in the middle age group, but the amount of the decrease is unknown. In animals a considerable variation of the LD-50 value with age has been observed, but for man there are hardly any data for the effect of age.

As mentioned in section 3.1.2, in addition to their greater sensitivity to radiation children may receive a larger dose to the marrow than would adults for the same surface tissue dose.

3.3.7. Acute lung effects

Death from the acute effects of radiation may also occur as a result of internal exposure following the inhalation or ingestion of radioactive substances, for example from fall-out. Inhalation is of particular importance in its effects on the lungs.

The inhalation of radioactive particles, if the dose to the lung tissue is high enough, may produce acute effects, even leading to death, quite apart from long-term effects such as fibrosis and cancer of the lung, which may result from much smaller doses. The acute effects ensue both from the direct action of the radiation on the lung walls and from the damaging action on the cells of the lung. The former affects the permeability of the membrane of the alveoli (air-sacs of the lungs), allowing fluids to escape into them. The symptoms of this are coughing, shortness of breath, and a feeling of 'drowning' in lung fluids. The effects on the lung cells include swelling of the alveoli, resulting in reduced gas exchange with subsequent hypoxia (deficiency of oxygen); haemorrhage into the alveolar spaces, giving rise to blood-stained sputum; loss of surface-acting secretion, leading to collapse of the alveoli and consolidation of the lung; and loss of the immunological function of the lung, leading to infection and pneumonia.

The causes of death may be heart failure due to hypoxia, pneumonia, or generalized toxaemia. The time of death may be some months after the inhalation; it depends on age, environmental conditions and availability of treatment. The lethal dose to the lung is about 10–20 Gy. But a person inhaling a radioactive material is also likely to have internal exposure to other organs, as well as external exposure. The combined, possibly synergistic action (see section 3.8) of the individual exposures makes the prognosis much worse. Under such circumstances smaller doses to the lung may prove lethal.

3.3.8. Other short-term effects

Apart from the immediate symptoms of radiation sickness, those who survive the exposure exhibit some transient symptoms. One of the earliest to occur is the effect on the skin.

Skin reactions were the first biological effect of radiation to be observed and reddening of the skin, erythema, was used in the earlier years of radiotherapy as a measure of the dose. The effects of radiation on the skin are somewhat similar to those from sunburn but extend much deeper. At a high enough dose erythema may develop within minutes and be accompanied by itching, flaking and pain. At lower doses symptoms may not appear until after about 10 days. For low-LET radiation erythema may appear after a single dose of about 3 Gy to the skin; a dose of about 6 Gy produces erythema in 50 per cent of those exposed. With neutrons, erythema starts at lower doses and

50 per cent may exhibit it after a dose of 2 Gy. The erythema is mild at small doses and gradually disappears. With doses above 10 Gy the skin lesions become progressively worse, as necrobiotic changes develop; the symptoms include scaling, blistering, weeping and sloughing. Ulcers often follow these changes, with permanent scarring and formation of keloids (ridges and nodules) of fibrous dense tissue; areas on the edge of the ulcer may develop cancer.

The permanent skin reactions occur after such high doses that if the whole body were exposed the person would die from the acute effects. But high doses to the skin alone can be received if beta-emitting radioactive substances from fall-out are deposited on the skin; such 'beta-burns' were observed among the inhabitants of the Marshall Islands after a test explosion (see section 4.2.3.1).

Another acute, but not lethal, effect of radiation is epilation (loss of hair) which starts about two weeks after exposure of the scalp to a dose of 3 Gy or more; recovery occurs in a few months, except for high doses to the skin when epilation may be permanent.

Irradiation of the reproductive organs can have temporary or permanent effects, but there is a big difference in this respect between males and females.

In men, depletion of sperm-forming cells results in loss of fertility. But if a sufficient number of sperm-forming cells are left they will begin to divide again and fertility may be restored. Temporary sterility may be caused by a dose as low as 0.3 Gy. The higher the dose the longer the time for recovery, for example, after a local dose of 6 Gy it may take more than 12 years. However, recovery does not occur in all cases; with doses greater than 2 Gy permanent sterility may result in a proportion of persons irradiated.

In women, radiation causes irreparable damage to the oocytes and consequently permanent sterility may result from exposure to a high dose of radiation. The effect depends on the age at irradiation. In women over the age of 40 a local dose to the ovary of 4 Gy may cause permanent sterility. A dose of 6 Gy is likely to cause permanent sterility at whatever age irradiation takes place [9].

3.4. Long-term somatic effects

Exposure to doses low enough not to produce any early symptoms of radiation sickness appears initially to have been without any harmful effects. Similarly, persons exposed to higher doses but who survive the acute effects give the impression of complete recovery even though this may take a considerable time in some cases. However, this impression is misleading. Even small doses of radiation damage a large number of cells. Much of this damage, particularly if caused by X-rays or gamma-rays, is repaired by the body, but a residual damage remains and this may eventually manifest itself in a variety of ways which, as a rule, are deleterious to the person exposed.

Latent damage to the somatic cells can lead to two types of detrimental effects, called stochastic and non-stochastic. Stochastic processes are those governed by the laws of probability; this means that they occur at random in an irradiated population. In this group of effects, the higher the dose the greater the frequency of occurrence of the given effect; however the severity of the effect does not depend on the dose. In the usual jargon it is an 'all or none' effect. With non-stochastic effects the reverse is true; every person exposed may suffer this type of injury, but the severity depends on the dose, being so small at low doses that the effect may not be detectable.

3.4.1. Induction of cancer

By far the most important of the stochastic effects, as far as somatic damage is concerned, is the carcinogenic action of radiation, that is, the induction of cancer. Since the early days of the discovery of X-rays and radioactivity it has been known that exposure to high doses of radiation can induce cancer in the organ exposed. Cancers were produced by the radiations from radioactive substances long before the discovery of radioactivity, but it took a long time to recognize the fact, for example, that the high incidence of cancer of the lung which had been known to occur among workers in uranium mines was due to the inhalation by the miners of a radioactive gas, radon.

In the early years it was thought that only large doses could cause cancer and that there was a threshold, a safe dose, below which no cancer would be induced. Very few radiobiologists still cling to the threshold hypothesis and even they concede that the threshold is much lower than was thought earlier; alternatively, some claim that there is a quasi-threshold for low-LET radiation. However, the majority of scientific opinion has abandoned the threshold idea. The consensus is that even the smallest dose of radiation may give rise to a small but finite increase in the incidence of cancer; however, there is a considerable difference of opinion about the magnitude of the carcinogenic effect of small doses.

The reason for the uncertainty is that the cancers induced by radiation do not differ in their symptoms and prognosis from the cancers that occur in non-irradiated persons, and which may be caused by other carcinogenic agents or may occur spontaneously. The only way it is possible to say that radiation has caused a given type of cancer is if the incidence of that type of cancer in the irradiated population is higher than in a population used as a control, that is, in a group of persons who are in all respects identical except that they have not been exposed to the radiation. With high doses the incidence of induced cancers is sufficiently raised to leave no doubt, but with small doses the increase may be so small compared with the control incidence that it is lost in the statistical fluctuations. For example, in order to obtain a meaningful answer to the question of how much exposure to 0.01 Gy (1 rad) increases the incidence of cancer, one would need a population of many millions in the irradiated as well as in the control group. Even with laboratory experiments an impracticably large number of animals would be required to obtain a statistically significant effect from a very small dose. The only way to calculate the carcinogenic effect of small doses is by extrapolation from observations made at high doses. However, extrapolation gives rise to considerable uncertainties. The two main reasons are (*a*) that the shape of the curve relating cancer incidence to radiation dose is unknown, and (*b*) that even at high doses the carcinogenic effect of radiation in man is not known accurately.

3.4.1.1. Dose–response relationship for cancer induction

Several theories have been put forward on the mechanism of cancer induction by radiation but none is sufficiently well established to gain general acceptance. Indeed, there may be no single mechanism applicable to all types of cancer, and the relationship between cancer induction and dose may be different for different cancers.

Figure 8 shows in a diagrammatic form several types of relationship at low and high doses; the low-dose range may be up to 0.1 Gy and the high-dose range from 1 Gy.

Figure 8. Types of dose–response relationships at low and high doses

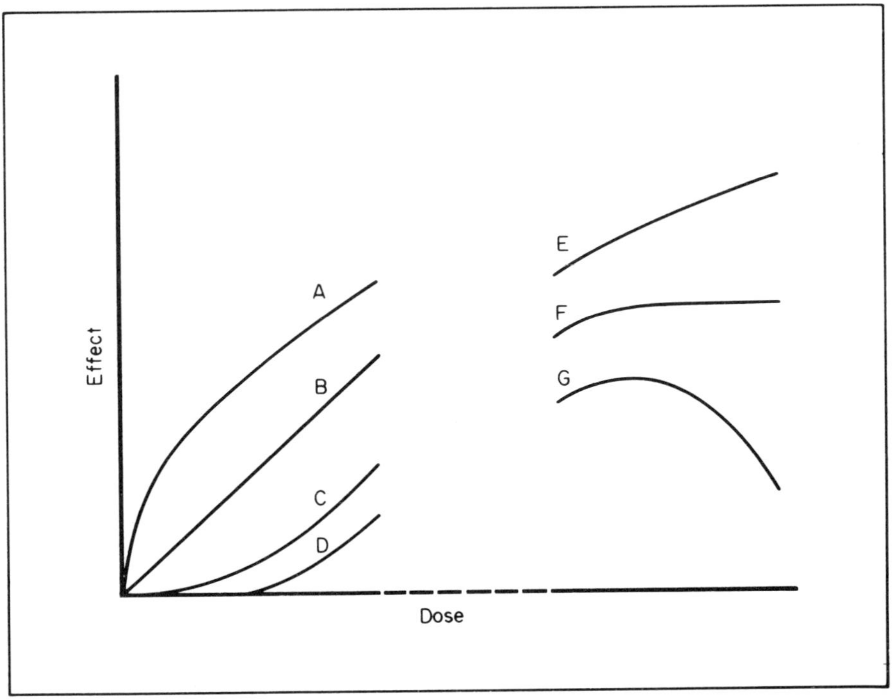

The simplest relationship is type B, namely a straight line. A linear relationship means that the increase in cancer incidence is proportional to the dose and starts from a zero dose. Type A is a supralinear relationship; for example, the increase in cancer incidence is proportional to the square root of the dose. With this type the rate of increase of cancer incidence (number of cancers per unit dose) is fastest at low doses and gradually decreases at higher doses. Type C is an infralinear relationship; for example, the increase in incidence is proportional to the square of the dose. In this case at low doses the increase is very small but the rate of increase becomes larger at high doses. Finally, type D represents a threshold relationship, with no cancer increase until a certain minimum dose, the threshold dose, is reached. As already mentioned, it is now accepted that this type of relationship does not apply to cancer incidence.

Before discussing the three curves in the high-dose region it should be emphasized that the curves E, F and G bear no relation to the curves A–D; any of the curves in the low-dose region could have the shape of any of the curves in the high-dose region.

In type E the incidence of cancer continues to increase, linearly or not, as the dose increases even to high values. In type F the incidence reaches a steady value, a plateau; this might occur, for example, if at a certain dose cancer was induced in all subjects exposed. In type G the cancer incidence, after reaching a maximum, decreases at higher doses. This seemingly strange effect has a simple explanation: high doses kill a large number of cells, including those which would otherwise have become cancerous. This decrease in cancer incidence is not of much comfort to the persons who receive such high doses, because they are most likely to suffer from other deleterious effects of high doses of radiation.

All the types of dose–effect relationship depicted in figure 8 have actually been observed in radiobiological experiments with animals. However, in relation to cancer induction different types of relationship were found to apply for different types of cancer and even for different conditions of irradiation, for example, different dose rates in the same type of cancer. Moreover, considerable variations have been observed between different species of animal. This means that animal experiments cannot be used for the quantitative assessment of the hazard to man, which can only be made on the basis of direct evidence from human exposure.

3.4.1.2. Induction of cancer in man

The past few decades have brought to light a number of observations of an increase in the incidence of various types of cancer following exposure to medium and high doses of radiation of human populations. The largest group are the survivors of Hiroshima and Nagasaki, whose experience comes nearest to that to be expected in a nuclear war as far as the initial radiation is concerned.

In relation to the effects of fall-out a small but important group are the inhabitants of the Marshall Islands in the Pacific, the region where large-scale testing of nuclear weapons was carried out by the USA in the 1950s. Indeed, the very first test of an F-F-F bomb on 1 March 1954 produced an intense fall-out in an unexpected direction. It came down on a Japanese fishing vessel, producing radiation exposure from which one of the crew subsequently died. It also caused a considerable exposure of the inhabitants of the Rongelap Atoll which resulted in a significant increase in cancer of the thyroid, as well as other cancers (see section 4.2.3.1).

Other important groups are patients who have been exposed to radiation for the treatment of a disease or who were X-rayed for the purpose of diagnosing a disease; later the groups were shown to have an increase in the incidence of certain cancers. Another group consists of workers who have received large doses of radiation over long periods of time in the course of their occupation, for example the miners in uranium or other mines with a radioactive atmosphere, as mentioned above.

The main purpose of the follow-up of the different groups of persons exposed to radiation is to establish a quantitative relationship between cancer incidence and radiation dose under various conditions of exposure and according to population characteristics. Such knowledge is of vital importance for setting safety regulations for both workers and populations in a world in which the use of radiation for peaceful purposes is steadily increasing. It is, however, also of importance for estimating the possible long-term casualties in a nuclear war, in regions remote from the combat areas, where the radiation exposure due to fall-out might be quite low.

However, the establishment of a dose–response relationship from these data has turned out to be very difficult for a variety of reasons.

1. The doses of radiation received by the persons exposed had in most cases not been measured accurately, if at all, and often the doses had to be guessed.

2. The conditions of exposure and the characteristics of the population differed enormously: in some the exposure was extremely short, while in others it was protracted over years; some groups received whole-body exposures, while in others only some organs were exposed; in some groups the population exposed consisted of young children, in others only of adults, and in still others only of women.

3. The observed incidence of cancer was often very small so the calculated frequency contains a large statistical error.

4. There is a latent period between the exposure and the appearance of the cancer, or more correctly the probability of the cancer appearing is very low soon after the exposure but gradually increases until a plateau is reached which may be of finite duration. Both the latent period and the plateau vary with the type of cancer, with the age at irradiation and with the dose. For leukaemia the latent period for adults is about 5 years and the plateau lasts for some 20 years, but for other cancers the latent period is about 15 years or longer, and the plateau may go on to the end of life, which means that an observed increase of cancers after 20–30 years may still underestimate the incidence.

5. The possible decrease of cancer incidence at high doses, due to cell killing (curve G of figure 8), makes it much more difficult to extrapolate from high doses to low doses.

6. Finally, and most importantly, a definite increase in cancer incidence has in most cases been established only at high doses, and as noted above the extrapolation to low doses is bound to be uncertain.

Figure 9. Incidence of leukaemia among survivors as a function of dose

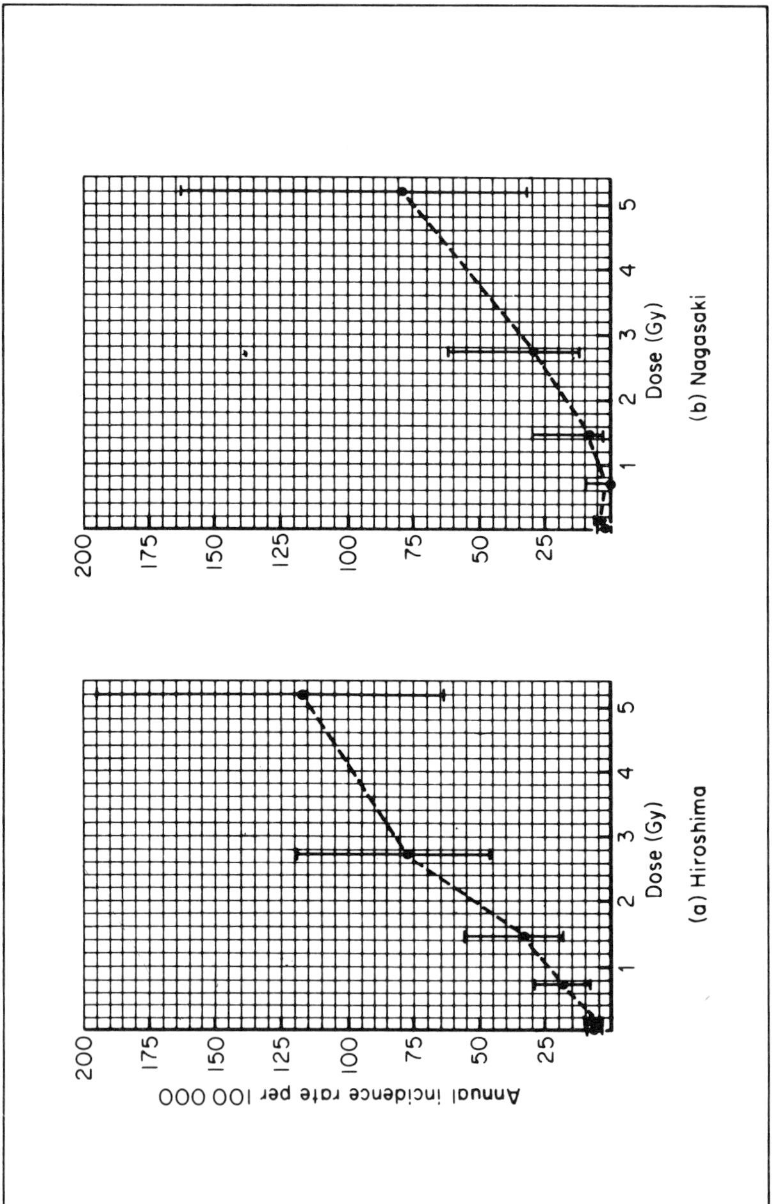

To sum up, the available data do not make clear which of the types of relationship shown in figure 8, or which combination of them, holds. The best quantitative material is provided by the Japanese survivors. As a result of painstaking efforts the dosimetry was thought to have been established fairly accurately; a large sample of the surviving population has been followed very carefully and all incidences of cancer have been recorded. Figure 9 shows the observed incidence rate of leukaemia during the period 1950–78, as a function of dose (*a*) for Hiroshima and (*b*) for Nagasaki [10]. As already noted, leukaemia is the earliest cancer to appear; by now the radiation-induced leukaemia appears to have passed the plateau period, and the incidence rate is getting near to normal. These are therefore the best data that can be obtained from human experience so far, yet the statistical errors (the bars in figure 9 represent 95 per cent confidence limits) are so large as to make it impossible to establish with confidence the nature of the relationship between incidence and dose.

An inkling of the degree of uncertainty and of disagreement among scientists on this issue is obtained from the history of the preparation of the Report of the Committee on the Biological Effects of Ionizing Radiations (so-called BEIR III) of the US National Academy of Sciences [5]. After the Report was completed and presented for publication the Academy withdrew it and requested a revision of the part dealing with the carcinogenic effect of radiation, in view of strong dissension by some members of the Committee. The final draft contains a compromise version, but attached to it are minority statements from the Chairman of the Committee and one of its members; one of these statements asserts that the formula adopted underestimated the risk, perhaps by an order of magnitude, while the other statement says that it overestimated it by a similar margin.

Of the various types of dose–effect relationship at low doses, the BEIR Report considered in detail the three depicted in figure 10. Type *a* is the linear relationship. Type *c* is the quadratic relationship (increase in cancer incidence proportional to the square of the dose). Type *b* is a combination of these two, a linear-quadratic relationship; this means that initially the cancer incidence increases in proportion to the dose, but the rate of increase is then faster, eventually becoming proportional to the square of the dose.

There seems to be general agreement that for high-LET radiation, for example for neutrons, the linear relationship is valid. This relationship appears to apply even for low-LET radiation, for example gamma-rays, for some types of cancer. But for leukaemia, exposure to gamma-rays seems to fit better the curves of type *b* or *c* rather than *a*. However, a comparison with the data in figure 9 shows that any of the three types could fit. Indeed, a detailed statistical analysis of leukaemia incidence in Nagasaki, where the radiation practically consisted only of gamma-rays, has shown that all three curves of figure 10 would equally well fit the observations.

To add to the confusion a recent study [11] has thrown considerable doubt on the validity of the dosimetry in the two Japanese cities. According to this study the neutron component in Hiroshima was much smaller than

Figure 10. Increased cancer incidence as a function of dose according to three models: (*a*) linear, (*b*) linear-quadratic, and (*c*) quadratic

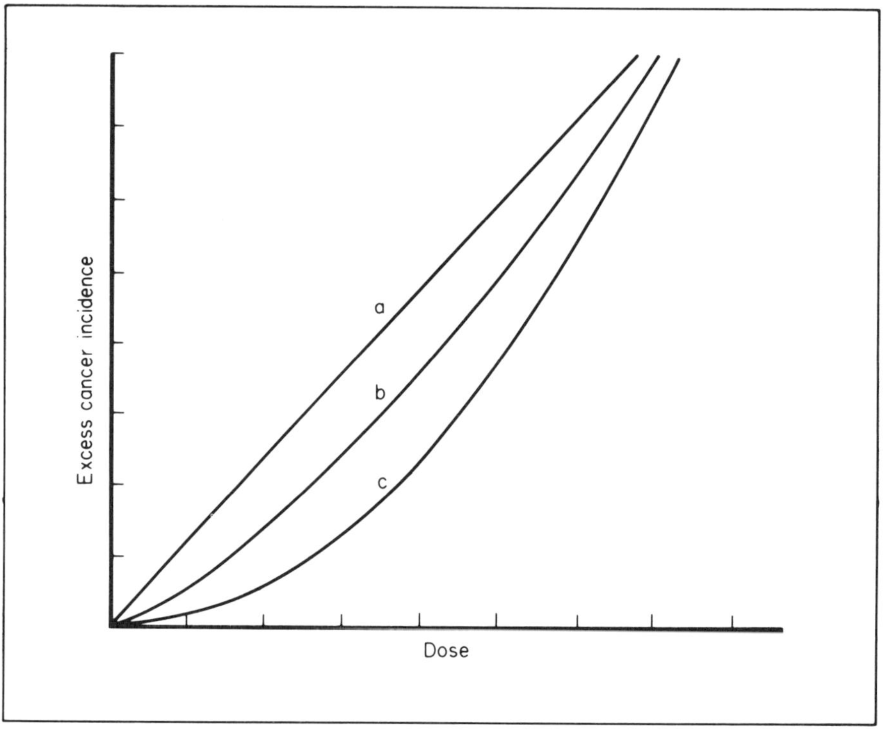

previously thought, the gamma-ray dose in that city should be increased and the gamma-ray dose in Nagasaki should be decreased. With these modifications, the difference in the rate of leukaemia incidence between the two cities (figure 9) would disappear and the linear relationship would again represent the best model.

It remains to be seen whether this criticism of the dosimetry is valid. In the meantime, the problem of how to extrapolate from high to low doses remains unsolved.

3.4.1.3. Risk factors for cancer

The protection of persons occupationally exposed to radiation requires regulations which set out the dose limits beyond which a worker should not be exposed in a given time. Similarly, regulations are needed about releases of radioactive materials from nuclear power or other industrial installations so as to keep the level of exposure of the general population within acceptable values. For the purpose of such regulations an assessment has to be made of the hazard from exposure to a given dose. The International Commission on Radiological Protection (ICRP) is the body which assesses the hazards and on this basis makes recommendations about exposure levels. These recommendations are generally accepted by governments in their radiation safety regulations.

In the absence of better evidence the ICRP has accepted the linear relationship as a working hypothesis for all types of radiation. Using the data available from the various groups of human populations exposed to radiation, as discussed in the previous section, the ICRP calculated the slope of the line for the carcinogenic effect of a given type of cancer. This slope gives the probability of the effect occurring per unit of dose and is called the risk factor. Since some cancers can be cured or are not always lethal one has to distinguish between the risk factors for cancer incidences and those leading to death. Table 9 shows the ICRP risk factors of death from the given type of cancer. They can be expressed as the ratio of the number of deaths from a given type of cancer to the total number of persons exposed to 1 sievert. No allowance was made for the age or sex of the person exposed. Thus, the risk factor for breast cancer is the average for both sexes, but since this cancer occurs almost exclusively in females it means that for them the risk is double the value given in the table. The other organs mentioned in the last line of the table refer to cancer of the stomach, lower large intestines, salivary glands and liver.

Table 9. Risk factors for cancer deaths

Cancer type	Death rate per sievert
Leukaemia	2.0×10^{-3}
Breast cancer	2.5×10^{-3}
Lung cancer	2.0×10^{-3}
Bone cancer	0.5×10^{-3}
Thyroid	0.5×10^{-3}
Other organs	5.0×10^{-3}

Source: International Commission on Radiological Protection Recommendations, ICRP No. 26 (Pergamon Press, Oxford, 1977).

The total risk comes to 12.5×10^{-3} per sievert; this means that if a population of one million persons is exposed to a dose of 1 sievert, 12 500 of them may be expected to die of cancer induced by the radiation. The actual number of cancers induced may be twice this figure to allow for the non-lethal cancers; in particular, cancer of the thyroid has a low mortality rate. But medical treatment is necessary to reduce mortality.

Apart from the controversy about the type of extrapolation to be used, the correctness of the ICRP risk factors themselves has been challenged. Some scientists contend that the slopes of the lines drawn by the ICRP on the basis of the linear model are too low by a large factor, perhaps as much as an order of magnitude. There is criticism of putting too much reliance on the observations of the Japanese survivors which—it is claimed—are a selected population and therefore not a representative sample. (This is quite apart from the doubts about the dosimetry.) There are also numerous, but not well-documented, reports of higher incidences of cancer than expected from the ICRP figures, among soldiers exposed to radiation in test explosions and among the populations in the vicinity of these tests. Similar claims were made in relation to persons exposed as a result of an accidental release of plutonium. The most comprehensive study of cancer incidence at low doses was carried out on workers in the Hanford nuclear plants; the reported results [12] suggest very much larger risk factors for some types of cancer, but the method of analysis used in that study has been heavily criticized.

In summary, there is no doubt that exposure to sub-lethal doses of radiation causes an increase in the incidence of most of the types of cancer that occur naturally, but the magnitude of the risk is uncertain, and there exists the possibility that at low doses the present estimate of the risk may be wrong by a factor of 10 in either direction.

3.4.2. Induction of cataracts

The most important of the non-stochastic effects is the effect on the eye.

The lens of the eye has a fairly high sensitivity to radiation and exposure of the eye may give rise to lens opacities. Radiation cataracts may range from small lesions, which cause only minor impairment of vision, to total blindness, the severity of the lesion increasing with increasing dose. With small doses the effect may be too small to affect vision and it is generally assumed that there is a threshold for the formation of a recognizable lesion; for low-LET radiation this threshold is about 2 Gy if the dose is given at a high rate. For neutrons the threshold is lower.

There is a latent period between exposure and the appearance of cataracts; depending on the dose and dose rate it may vary from 1 year to 10 years. Again, depending on the magnitude of the dose the cataract may regress with time, remain stationary, or become worse.

3.4.3. Life-shortening

Another somatic effect is shortening of the life-span after whole-body irradiation. Such an effect was definitely established in animals, where a linear dose–response relationship was found for single exposures. This amounted to a reduction of the life-span by about 5 per cent per gray. The life-shortening was thought to be the result not only of the earlier occurrence of cancers but also of an increased rate of non-malignant diseases, and could be considered as a non-specific effect suggesting premature ageing. In man, evidence of life-shortening was observed among radiologists in the United States who used to receive high doses of radiation as an incidental by-product of their occupation. Their mortality rates were significantly higher than those among other medical practitioners and resulted in a reduction of the life-span by about five years.

No symptoms of premature ageing or of life-shortening have been observed among the A-bomb survivors in the Japanese cities. The mortality rate from non-malignant diseases does not appear to be significantly different from that in the non-irradiated population.

3.4.4. Effects of pre-natal exposure

The human embryo and foetus are highly sensitive to radiation and a number of deleterious effects were observed in children born to mothers who were exposed to ionizing radiation during pregnancy, at doses much lower than might have been expected from observations on adults.

Among the women who were pregnant during the explosions in Hiroshima and Nagasaki there was a marked increase in still-births, and the mortality of their children during the first year of life was considerably increased. The surviving children showed a greater frequency of mental retardation and they had head circumferences which were smaller than normal; other malformations included defects of the skeleton and the eyes. The stage of gestation at the time of exposure influences the type and frequency of the impairment. The risk of malformation is largest in the early stages of pregnancy but the risk of pre-natal death increases in the later stages.

Exposure in pregnancy also increases the rate of induction of cancer in children born to these mothers. Even a diagnostic dose of about 10–20 milligrays may increase the incidence of leukaemia and other cancers by some 50 per cent. These cases of leukaemia occur between three and eight years of age, with a peak at about five years. It is not yet known whether the incidence of leukaemia will rise again in adult life. The absence of an increase in cancer rate among children born to women in Hiroshima and Nagasaki who were pregnant during the explosions is one of the odd features of these survivors.

3.5. *Genetic effects*

If the germ cells of a person—as distinct from the somatic cells—receive a dose of ionizing radiation, changes may occur which would manifest themselves in the offspring of the exposed person or in future generations; these are referred to as genetic effects. Two types of damage may occur, gene mutations—that is, an alteration in the structure of one of the genes—and chromosome aberrations, which may affect a number of genes at the same time. It is generally accepted that both types of change are harmful to the descendants, but the type and magnitude of the effects vary enormously, from barely detectable to lethal in their consequence. In the latter case, if the result of the genetic change is death *in utero* or in early life, this particular type of defect will be removed from the genetic pool of the population. But non-lethal defects will be carried from generation to generation, particularly when the welfare society enables affected individuals to survive to maturity and beget children. Some of the changes, known as recessive mutations, reveal themselves only when both parents happen to have had the same mutation. These mutations may thus be carried through a number of generations before the effect of the irradiation becomes manifest.

Any genetic defect that may result from exposure to radiation is likely to be the same as one produced by other mutagenic agents and which already exists in the population. About 10 per cent of all live births carry a significant genetic defect and exposure to radiation may be expected to add to this burden an amount dependent on the dose. The genetic effect of radiation is often expressed in terms of the 'doubling dose', that is, the dose of radiation which would add to the genetic pool the same number of defects as occur naturally. The concept of a constant doubling dose implies linearity between effect and dose. However, the presentation of the genetic effect by a single linear relationship can only be an approximation. Experiments with mice have shown that different genes often differ in their sensitivity to radiation. There is also a dependence on dose rate which is different for males and females. The doubling dose would thus be a value averaged over many conditions.

Even with this qualification there is a large uncertainty in the value of the doubling dose. For man it is taken to be between 0.5 and 2.5 Gy, but it may well lie outside these limits. The reason for this large margin of error is that the doubling dose has to be extrapolated from data on animals, mainly mice, as there are no human data available.

The only irradiated population which was thought to be large enough to show genetic effects are the survivors of the A-bombs in Japan. But despite a prolonged investigation of the first generation of children born to the survivors no increase in the incidence of genetic damage has been established. The effects studied included the incidence of abortions, still-births, congenital defects, infant mortality, sex ratio of children and birth weight, as well as direct measurements of the incidence of chromosome aberrations. In none of these criteria was the difference between the exposed and control populations statistically significant.

Different interpretations of this negative finding were put forward: (*a*) that the methodology used in the investigations was faulty; (*b*) that the sensitivity of the criteria used was too low for the size of the population investigated; (*c*) that the survivors are not a typical population as far as radiation effects are concerned; (*d*) that a genetic effect does exist but with a lower frequency than deduced from animal experiments; and (*e*) that there is no genetic effect of radiation in man.

This last interpretation cannot be taken seriously, but it is being used in popular publications to convince the public that there is no reason to worry about genetic effects of radiation. Genetic effects of radiation have been observed in animal species and there is no reason why the human species should be unique in this respect. Nevertheless, the absence of any effect is puzzling. The ICRP has used a risk factor for genetic damage in the first two generations of 4×10^{-3} per sievert; this was based partly on animal data and partly on the lower limit of the doubling dose that can be deduced from the negative findings in Hiroshima and Nagasaki. But like the somatic risk factors, the ICRP figure for genetic damage may be wrong by a large factor.

3.6. Factors affecting the biological response to radiation

As already intimated, the sensitivity of biological systems to radiation, even for a specific effect, for example the probability per unit of dose of inducing a given type of cancer, is greatly influenced by many factors. Some of these are of a physical nature, such as the quality of the radiation, the dose rate, or its spatial distribution, while others are of a biological nature, such as age, sex and the effects in individual organs.

These factors are discussed in the following sections, but it should be pointed out that, in addition to these factors, there appears to be a variation in the response to exposure to radiation which cannot be ascribed to any specific factor. It is quite likely that the genetic make-up of a person plays an important role in the intrinsic sensitivity to radiation in regard to both acute and long-term effects. The human species is genetically very heterogenous and if certain genotypes do confer a greater or lesser susceptibility to radiation, then large and apparently random differences in biological effect may be found between individuals, or groups of populations, exposed to the same dose.

3.6.1. Radiation quality

The different types of radiation accompanying nuclear warfare are emitted with a spectrum of energies which ranges over several orders of magnitude and whose biological effects may vary considerably. The physical property of the radiation that affects its biological effectiveness is LET (section 3.1.1), which is a measure of the density of ions, directly or indirectly produced by the radiation. The ion density depends on the charge and velocity of the ionizing particle. The higher the charge, the greater the ion density; for a given charge, particles with a low velocity produce more ions per unit length than do faster particles. Thus, alpha-particles ionize very intensely because they carry two electric charges and have a relatively low velocity for a given energy. (The mass of an alpha-particle is about 7 000 times greater than the mass of the electron and its velocity is therefore about 85 times smaller than that of an electron of the same energy.) Neutrons, which ionize indirectly, mainly by setting protons into motion, are also classified as high-LET radiation. By contrast, beta-rays are poor ionizing agents. X-rays and gamma-rays, which ionize indirectly via the electrons they produce, are also in the low-LET class.

Although there is no general theory which can adequately explain all forms of radiation damage to cells and tissues, it is possible to understand why the ion density of the radiation has such an important bearing on the efficiency of cell killing. One major cause of cell death is damage to the chromosomes—the threadlike structures which carry the genetic material and whose integrity is essential to the viability of the cell. These chromosome threads are for most of the time two-ply, and both strands must be damaged in closely adjacent positions if the thread is to be broken irrevocably. Single-strand breaks are rapidly repaired by the cell's own enzymatic defence mechanisms but double-strand breaks are not easily repaired.

The destructive energy concentrated along the track of an ionizing particle of high LET is usually sufficient to break both strands in a single passage through the chromosome, whereas a particle of low LET may often cross the thread without leaving enough energy to break even a single strand, and it will very rarely break both strands simultaneously. Two or more separate low-LET particles must therefore cross the chromosome thread at closely adjacent positions to produce a double break and this will generally require a considerably higher average dose than that for a single transit by a high-LET particle.

The relative biological effectiveness (RBE), or quality factor, is a measure of the LET effect in any specified type of biological damage. When the biological effects of different types of radiation are to be assessed, the dose equivalent (in sieverts), which is obtained by multiplying the dose (in grays) by the RBE, should be used. By definition, the RBE of low-LET radiation (X-rays or fast electrons) is unity.

However, for a given biological end-point the RBE is not constant but may vary with the magnitude of the dose delivered. This occurs particularly when the high-LET radiation produces 'overkill', that is, when a significant proportion of the sensitive structures receive several hits from high-LET particles where one alone would suffice to inactivate them. For this reason the RBE of fast neutrons is nearly unity at very high doses.

As the dose is reduced the RBE for neutrons gradually increases but it should reach a constant value at very low doses. The reason for this is that near the end of the range, when the velocity is low, even electrons behave like a high-LET radiation. There is a high-LET component in any radiation and it becomes dominant at low doses. This is one of the reasons for the linear-quadratic dependence on dose that was assumed for cancer induction by gamma-rays (see section 3.4.1.2).

For the neutrons which occur in nuclear warfare an average quality factor of 10 is adopted at dose levels up to a few grays. At very small doses it may go up to 50, whereas at high doses it decreases to unity (but see section 3.3.6).

For alpha-particles the average value of the quality factor is taken as 20.

3.6.2. Dose rate and fractionation

The effectiveness of a given dose of low-LET radiation is expected to be smaller if the same dose is spread out over a longer period of time or is given in a number of separate fractions. The reason is the same as explained in the previous section. A given dose delivered at a low rate, or in small portions, provides a greater opportunity for the damage done by the earlier portion of the dose to be repaired. For acute effects the dose-rate dependence is well established (see section 3.3.6). This fact is made use of in radiotherapy, where large doses have to be given which might kill the patient if delivered at once. But for long-term damage the dose-rate effect is much less evident. The carcinogenic effect observed in human populations came from exposures given at a very wide range of doses, from a practically instantaneous exposure as in Japan, to chronic irradiation as in the case of uranium miners. Yet very little difference in the risk factor was noted and no guidance is provided by the available data about the influence of dose rate on long-term effects. For high-LET radiation there appears to be a reverse dose-rate effect: the probability of induction of some cancers is greater if the dose is delivered at a lower rate.

The ICRP assumption of a linear relationship implies that there is no dose rate dependence for long term effects. At the present state of knowledge this is probably the prudent assumption to make.

3.6.3. Partial-body exposure

Many effects of radiation are greatly reduced if only part of the body is exposed. A dose of radiation which would be lethal if given to the whole body will often produce no mortality if given to part of the body. However, the particular organ irradiated may exhibit a serious radiation reaction, sometimes developing into a malignancy which could be fatal. For long-term effects too, partial-body irradiation may be less damaging. Thus, the induction of leukaemia is greatly reduced if only part of the bone marrow is exposed to radiation. For other cancers, where local irradiation involves the whole organ, such as the breast or the thyroid, there is little difference from whole-body irradiation.

Alpha-particles, which have a very short range in tissue (up to about 50 micrometres), produce a very localized exposure. Moreover, alpha-emitting nuclides are often selectively concentrated in thin layers of an organ (e.g., at the growing ends of bones) and these thin layers therefore receive a very high dose and are strongly affected. To allow for this higher risk a special distribution factor is introduced when calculating the biological effects of internally deposited alpha-emitters.

3.6.4. Effect of age

Apart from the high radiosensitivity of the foetus (see section 3.4.4), there appears to be a variation of sensitivity to radiation with the age at exposure. Young children are the most vulnerable; with increasing age the sensitivity is reduced, reaching a flat minimum in early adulthood and then rising slowly again for older age groups. The available data are not accurate enough to make possible a quantitative estimate of the variation of radiosensitivity with age, but it is likely that in early childhood the sensitivity is several times higher than for adults.

The risk factors for induction of cancer are also affected by the age at exposure, but the age dependence is not the same for different cancers. From observations on the Japanese survivors it appears that in children there is a higher incidence of leukaemia per unit dose; it starts earlier and the plateau is of shorter duration than in adults.

The incidence of thyroid cancers among the children in the Marshall Islands following the intake of radioactive iodine (see section 4.2.3.1) was significantly higher than in adults. This may be partly due to the fact that children have smaller glands and drink more milk, which is the main source of radioactive iodine.

The incidence of cancers of the breast and lung also appears to be dependent on age at irradiation, but it differs for these two types of cancer and is not well established.

3.6.5. Effect of sex

The risk factors for the induction of cancer are sex-dependent for some cancers but not for others. Cancer of the breast and thyroid appear to be among the most sensitive to radiation and since these cancers occur more frequently in women than in men the overall carcinogenic risk factor for women is probably about twice that for men.

There is a marked sex difference in the sterility effect (see section 3.3.8). With regard to genetic damage, a sex difference in the variation of the genetic effect with dose rate was observed in animals. At lower dose rates the frequency of mutations per unit dose is much smaller in females than in males. However, as stated above, no genetic effect of radiation has so far been observed in man.

3.7. Internal exposure

The fall-out from nuclear weapons, apart from giving rise to external irradiation of the human body, also produces internal irradiation from the radioactive materials ingested or inhaled. Depending on their chemical and biological properties, these materials often concentrate in certain organs and remain there for longer or shorter periods. During their stay in the body they deliver a dose of radiation the effect of which is similar to that of external irradiation, except that the exposure is very unevenly distributed. Thus, the same total amount of radioactivity which would produce a low dose if uniformly distributed over the body may give a large dose to a particular organ in which it has become concentrated. For many radioactive nuclides a critical organ is defined which is the most vulnerable to the particular nuclide. For example, for radioactive iodine the thyroid is the critical organ, as iodine concentrates in the thyroid to a much greater extent than in the rest of the body.

The dose delivered to a given organ due to the presence in it of a radioactive nuclide depends on several factors, for example the amount of radioactive substance taken into the organ, the type and energy of the emitted radiation, and the length of time the radioactivity remains in the organ. The latter depends on the physical decay of the radioactivity (physical half-life) as well as on the rate of removal of the substance by normal metabolic processes. This rate is usually expressed in terms of the biological half-life, that is, the time in which half the amount of material present in the organ is removed. The biological half-life of a given radioactive element may differ, depending on the chemical form of the substance inhaled or ingested, for example whether it is soluble or insoluble.

For all these reasons the calculation of the dose resulting from the intake of a radioactive material is much more complex than in the case of external exposure. This applies in particular to alpha-particle emitters such as plutonium and other actinides.

The ranges of the alpha-particles are so short that they cannot be detected from the outside; indirect methods of detection have to be used but with present day techniques the results are not reliable. For this reason the estimate of the dose is usually based on theoretical models of the metabolism of the given substance in the body. In these models the body is divided into several compartments and rates of transfer of the radioactive substance from one compartment to another are assumed. Apart from the chemical form of the substance these rates depend strongly on the size of the radioactive particles. However, huge variations in the relative amounts transferred between organs occur from person to person, so that in practice the models are not of much use in calculating the internal dose from alpha-particles in any one individual.

In the case of plutonium, its deposition in the lung may give rise to such intense local irradiation that very small amounts of plutonium have to be considered hazardous to a high degree. From experiments with dogs it was deduced that 16 mg of insoluble plutonium in the lungs would cause death within one month (see section 3.3.7); 0.3 mg lodged in the lung would give a 50 per cent probability of cancer induction.

Although many studies have been made on the metabolism of various radioactive isotopes in different forms and on the doses that may result from a given concentration of them in the body, there remains much uncertainty about the damage that may result from them. In particular, the carcinogenic action of plutonium remains the subject of considerable controversy.

3.8. Synergistic effects

The estimates of the biological effects of exposure to radiation are assumed to apply to individuals who are otherwise uninjured. The situation is much worse if the person exposed has also been subject to other traumas, burns or mechanical injuries, as is likely to occur in a nuclear war. The outcome of combined injuries could be either additive or synergistic (it is most unlikely to be antagonistic, i.e., that the combination of injuries would give a better prognosis than each injury separately). An additive effect occurs when the outcomes of the individual injuries are independent of each other and the total effect is the sum of the separate effects. Synergistic action is one which potentiates the effect of each separate injury so that the overall effect is greater than the sum of the individual effects.

There are not many data about the result of combining radiation injuries with other traumas, but the evidence available suggests a synergistic rather than an additive effect. In animal experiments a combination of radiation with mechanical injury, or with burns, produced much greater mortality than was expected from the separate injuries added together. There is no reason why the same should not apply to man. Thus, the human body would be much less able to cope with a thermal injury if the immune mechanism and the bone marrow cells were depressed by the action of radiation. Similarly, damage to the skin and soft tissue by mechanical injury could lead to infection, and any blood loss could lead to serious consequences if the number of white blood cells and platelets were reduced by radiation.

For these reasons the chances of recovery of persons who receive a dose of radiation and also suffer other injuries would be considerably smaller than estimated from radiation exposure alone. In most calculations of the casualties in a nuclear war it is assumed that each of the three causes of injury—blast, heat and radiation—acts as if it were alone. Although it is impossible with the present lack of knowledge to give a quantitative estimate of the synergistic effect of two or three causes acting together, there can be no doubt that the total number of casualties would be considerably larger if this were taken into account.

The following imaginary example may help to explain numerically the effect of synergism. Suppose that as a result of a nuclear explosion a person received a thermal injury as well as a total-body dose of radiation. Suppose further that the burn alone—in the absence of other traumas—gave a 60 per cent chance of recovery and that radiation exposure alone gave a 70 per cent chance of survival. Then if there were no interactions between the two injuries the chance of survival would have been 42 per cent (the product of 70 per cent and 60 per cent). Now suppose that as a result of the exposure to radiation the chance of recovery from the burn went down to 30 per cent; and, similarly, for an injured person the radiation mortality curve of figure 7 (page 35) shifted to the left and the chance of survival decreased to 50 per cent. Then the total chance of survival would be only 15 per cent, which is about one-third of what it would have been without synergism.

A similar argument applies to the effects of a combined exposure to external and internal sources of radiation or to the internal irradiation of several organs. In the fall-out areas after a nuclear bomb explosion a person may inhale and ingest radioactive nuclides from the cloud, so that several organs (for example, the lungs and the gastro-intestinal tract) receive doses of radiation. At the same time the person may receive a total-body external dose from the gamma-rays emitted by the radioactive materials deposited on the ground. The interaction between the injurious effects of each of these events is likely to be such that the chance of survival would be much less than the sum of the chances of survival calculated for each exposure separately.

Not only physical injury but also psychological stress might conceivably adversely affect the prognosis of a person who received an acute dose of radiation, quite apart from the societal upheavals and the breakdown of medical services which are bound to accompany a nuclear war.

It is more difficult to envisage a synergistic influence on the long-term effects of radiation, but if the probability of cancer induction is very high, similar considerations would apply.

3.9. Protection against radiation effects

The most effective protection against radiation is not to receive it. In other words, the use of physical methods to absorb the radiation before it reaches the body is the surest way to reduce the risk. Shielding provided by walls of buildings or by an underground shelter can greatly reduce the dose of radiation; even if only part of the body were shielded, the prognosis would be much better (see section 3.3.6). Such physical methods of protection relate to civil defence measures and will be discussed in section 6.4.

Chemical or biological methods of protection are much less effective. A number of drugs, mainly sulphydryl compounds (cysteine, cysteamine), have been developed which reduce somewhat the severity of acute doses of radiation, but their effectiveness is of short duration and they have to be taken soon after the exposure or preferably immediately before. Moreover, they are generally toxic.

Injection of bone marrow is probably the most effective way to counteract the acute effects of whole-body exposure to doses up to 10 Gy. To ensure that the transplant will not be rejected, the recipient's own bone marrow has to be destroyed. Banks of bone marrow have been set up in some countries but the quantities are so small that this could not be applied on a large scale. In any case, such treatment requires intensive medical care and protection against infection and these facilities are unlikely to be available in the aftermath of a nuclear war.

For radioactive isotopes taken into the body from the fall-out, some chelating agents were developed which—it was claimed—acted to remove the radioactivity from the body faster. Some aspects of this finding are now in doubt; in any case, here too the scope of action is limited. The only effective and easily applied protective measure is that against damage to the thyroid by radioactive iodine. By taking tablets containing ordinary iodine the amount of radioactive iodine reaching the thyroid gland is greatly reduced. The radioactive iodine then goes into the blood stream from where it is soon removed by urinary excretion. But because of the fast turn-over of iodine, potassium iodide tablets would have to be taken over a period of time; they may also produce adverse effects in some people.

As far as hereditary and congenital effects are concerned nothing can be done at present, except to prevent the birth of children who are likely to be severely damaged. This entails abortions for pregnant women who receive a high dose, particularly soon after conception. Methods of 'typing' abnormalities in the chromosomes have been developed and are being made more sensitive. Persons in whom recognizable inheritable damage has been found would have to be persuaded not to have children, but a massive chromosomal assay of the population is beyond present capabilities.

4. Radiations from nuclear explosions

In a war involving the use of strategic nuclear warheads the weapons are likely to be employed primarily in the same way as any other high explosive, to cause the maximum destruction by blast. Nevertheless, the fact that the accompanying radiation may contribute considerably to the casualty toll and in many cases may exceed that caused by blast and heat will undoubtedly play an important part in the overall strategy.

The radiations accompanying or following a nuclear explosion are usually divided into two groups: initial and residual. Actually, the emission of radiation goes on continually from the moment of the explosion, but the origin of the radiation emitted immediately after the initiation of the explosion is different from that of the later radiation, and its rate of emission is much greater. It is conventional to use the term 'initial' for all radiations emitted during the first minute after the detonation. The reason for this convention is that after one minute the fire-ball has risen to such a great height that only a negligible fraction of the emitted radiation will reach the ground. The initial radiation consists mainly of neutrons and gamma-rays. The radiation emitted after the first minute is called 'residual' and most of it comes from the fall-out. It consists mainly of beta-rays and gamma-rays from the radioactive materials produced as a result of fission; but it also includes radioactivity induced by neutrons, and tritium from fusion bombs. Some alpha-particle emitters are also present in the fall-out, from the debris of the unspent fissile materials, uranium and plutonium, and from their products.

4.1. Initial radiation

4.1.1. Neutrons

Neutrons are emitted at both fission and fusion reactions. Therefore, the total number of neutrons produced at the instant of the explosion is proportional to the total energy released, whether it comes from a fission, a fusion or an F-F-F bomb. However, in any of the types of bomb the interactions between the neutrons and the materials of the bomb itself considerably modify both the number and the energy distribution of the neutrons emitted after the explosion. Thus, depending on the method of assembly of the bomb, the radiation hazards due to neutrons vary enormously, even for bombs of the same type.

For example, the two fission bombs detonated over Japan were thought to differ greatly in their neutron emission. In the Nagasaki bomb, which was detonated by the implosion technique, large amounts of hydrogenous material, mainly in the plastic explosive charges which initiated the implosion, surrounded the plutonium. Hydrogen is very effective in slowing down fast neutrons and in absorbing slow neutrons. The result was that there was hardly any neutron component in the radiation emitted from the Nagasaki bomb; the exposure of persons was almost entirely to gamma-rays. By contrast, in Hiroshima, where the gun assembly was used, the iron of the gun attenuated the gamma-rays more than the neutrons; consequently, the latter were believed to have contributed a significant proportion (about 20 per cent) to the surface tissue dose received by the population, but this has been challenged (see section 3.4.1.2). Due to the high RBE of the neutrons (see section 3.6.1) the long-term radiation effects among the Hiroshima survivors, in the low dose groups, were believed to have been caused much more by neutrons than by gamma-rays.

Fission and fusion bombs differ in the initial number of neutrons produced per unit of explosive power as well as in their energy distribution. For a given explosive yield the number of neutrons produced in fusion reactions is about 10 times greater than in fission reactions. In the latter the neutrons are emitted with a continuous spectrum of energies (see figure 11) ranging from nearly zero up to more than 1 pJ, but with most of the neutrons having energies around 0.1 pJ, at which value there is a peak in the spectrum. By contrast, the neutrons of the fusion reactions (see table 2, page 7) are emitted with discrete and much higher energy values; for example, the neutrons from the d–t reaction are all emitted with an energy of 2.8 pJ.

Figure 11. Energy spectrum for fission neutrons

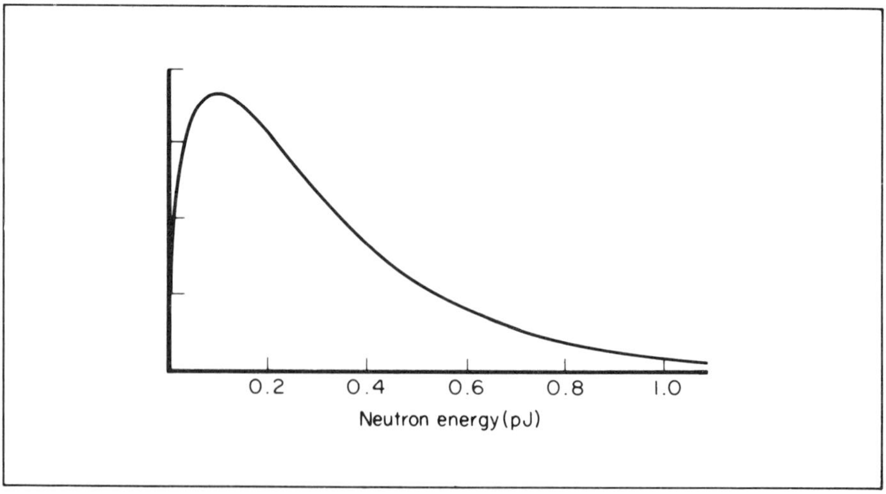

However, as indicated, immediately after their production and before the bomb breaks up, the neutrons make numerous collisions with the atomic nuclei of the materials of the bomb; in such collisions most of the neutrons are captured and those remaining lose energy to a varying degree. The result is that with both fission and fusion bombs there is a continuous spectrum of neutron energies, although still with a larger component of high-energy neutrons in fusion than in fission bombs. In both cases the mean energy is considerably lower than in the original spectrum.

Depending on the amount and type of material used in the assembly of the bomb, the total number of neutrons which are emitted after the explosion may vary by several orders of magnitude for bombs of the same explosive yield. There is thus a very wide range of neutron doses which may be received by persons at a given distance from the explosion, even after the detonation of a bomb of a specified yield. In the following, certain values of the neutron emission per kiloton of explosive yield (approximately 10 per cent of the original number of neutrons) are assumed as representative values, but it should be remembered that the neutron emission can be made much smaller or much larger, depending on the type of damage it is intended to inflict. In particular, bombs can be designed so that most of the damage is due to neutrons (as is the case in the neutron bomb, section 4.1.4). However, in the strategic use of nuclear weapons the main emphasis is likely to be on the mechanical and heat effects. According to convention we shall assume that thermonuclear weapons would be of the F-F-F type, with half of the total energy yield coming from fission and half from fusion.

As the neutrons travelled away from the source they would spread out over a large area so that the fluence, that is, the number of neutrons falling on a unit of surface area, would decrease with distance. If the neutrons were travelling in vacuum this decrease would follow an inverse square law; for example, the fluence at a distance of 2 km would be four times less than at 1 km. However, the reduction of fluence due to the geometrical factor is small compared with that due to the interaction of the neutrons with the nuclei of the atoms through which they pass. As they travel through the air, the neutrons undergo an attenuation in number and in energy. The absorption of neutrons follows an exponential law; this means that layers of air of equal thickness reduce the intensity by equal proportions. The rate of absorption depends on the density of the air and therefore on the height above ground of the explosion, since air density decreases with altitude. At a given slant distance from the explosion the neutron fluence would be greater the higher the point of detonation, because a larger proportion of the neutron path would then be through rarified air. If we assume an average air density of 1.16 kg m^{-3} (which is 0.9 of the normal sea level density), the fluence, for an average energy of the neutrons, would decrease by a factor of 10 each time the neutrons passed through a layer of air 570 metres thick; this may be taken to apply to neutrons from both fission and fusion bombs. This rate of reduction in fluence is in addition to that due to the inverse square law.

When the neutrons reach the human body they are partially absorbed in it and in this way deliver a dose of radiation. For the same neutron fluence the dose delivered varies with the energy of the neutrons; therefore, to determine the dose at a given distance calculations have to be made which take into account not only the number of neutrons but also the spectrum of energies as it varies with distance. The result of such calculations is presented in figure 12, which gives the neutron dose (in tissue at the surface of the body) as a function of the slant distance for a 1-kt bomb; the same curves can be used for fission and F-F-F bombs. The exponential attenuation of the dose is greater than that of the fluence; an increase in slant distance by 510 metres reduces the dose by a factor of 10.

Figure 12. Neutron dose (surface tissue) per kiloton as a function of slant distance

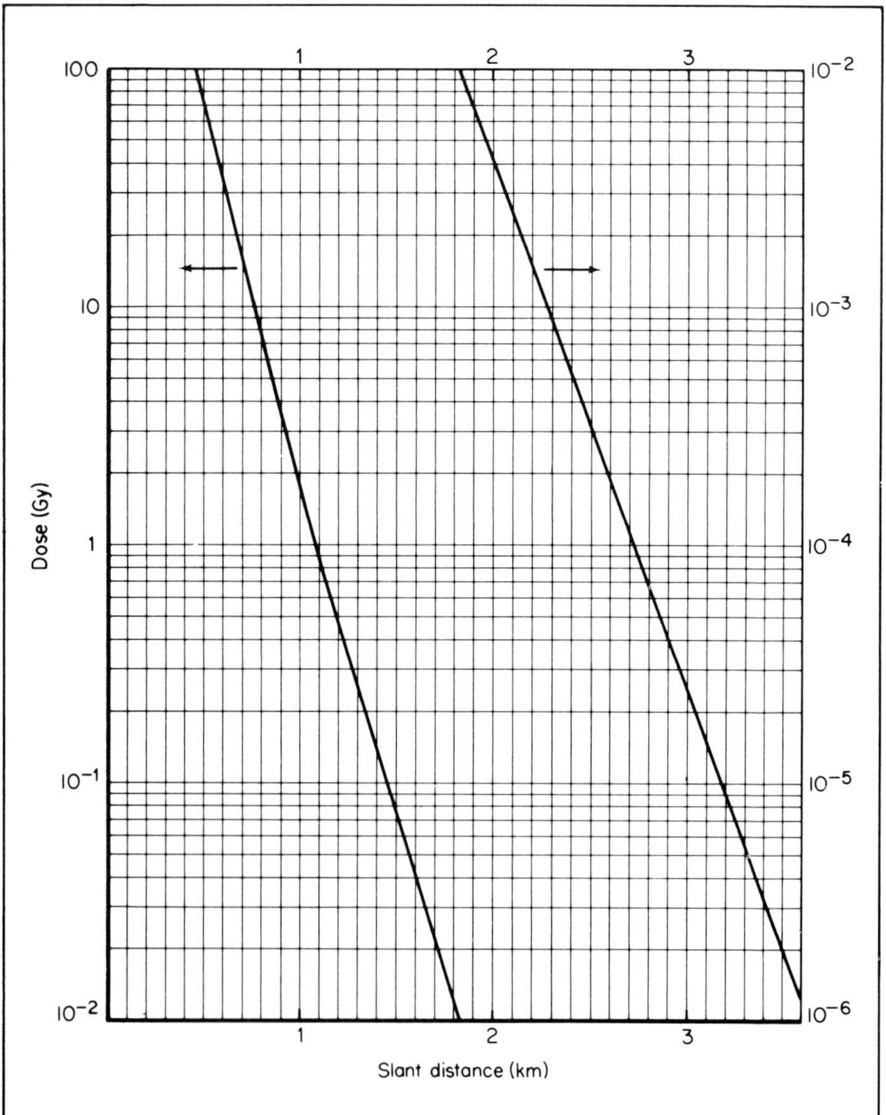

The curves of figure 12 apply to an air burst at a height greater than 100 metres. For a surface burst the dose values should be multiplied by 0.5 (because half of the neutrons are absorbed in the ground); for intermediate heights a proportionality factor between 0.5 and 1 needs to be applied.

Since for a given type of bomb assembly the number of neutrons is proportional to the explosive yield, the neutron dose for a bomb of any other yield can be read from these curves simply by multiplying by the explosive yield. Thus, at a slant distance of 1 km, a 1-Mt thermonuclear bomb would deliver a dose of about 2 000 Gy; at 2 km the dose would be about 4 Gy.

Figure 12 illustrates the rapid decrease of the neutron dose with distance due to the exponential absorption; it is a much faster decrease than for mechanical or thermal damage (see section 4.1.3).

It should be emphasized again that the doses in figure 12 are values calculated for an assumed design of the weapon and for a specified air density. Under other conditions the neutron doses may differ by a large factor in either direction.

4.1.2. Initial gamma-rays

The gamma-rays emitted during the first minute after the explosion originate from several nuclear processes. For fission bombs the most important of these processes are listed in table 10. As is seen, three of the gamma-ray sources (2, 3 and 4) are of secondary origin and result from the interaction of neutrons with nuclei of weapon materials and of air. These three sources also occur in fusion bombs.

Initially, the prompt gamma-rays (source 1) have the greatest intensity, but most of them are absorbed in the weapon material and less than 1 per cent emerge. Further absorption takes place as the gamma-rays pass through the air, the gamma-rays of lower energy being attenuated with distance faster than those of higher energy. With both fission and fusion bombs source 4 gradually becomes predominant, because its gamma-rays have a higher energy than those from the other sources.

Table 10. Initial gamma-rays from nuclear explosions

Source	Energy (pJ)	Time of emission
1. Prompt gamma-rays from capture of neutrons by uranium or plutonium	~0.2 (an average of 7 gamma-rays per fission of total energy 1.2 pJ)	0.05 μs
2. Inelastic scattering of neutrons by nuclei of weapon materials	~0.1	~1 μs
3. Inelastic scattering of neutrons by nuclei of air	~0.2	~10 μs
4. Radiative capture of neutrons by nitrogen of the air	~1	~10 ms
5. Decay of fission products	0.02–0.4 (total of 1.1 pJ per fission)	continuous

For F-F-F bombs of high yield another phenomenon related to air density makes source 5 predominant at large distances. This is the 'hydrodynamic enhancement' of the gamma-ray dose and arises from the continuous time of emission of the gamma-rays from the fission products. The gamma-rays from the other sources are all emitted before the shock wave, and travel ahead of it in air which is still at normal density. However, the gamma-rays from the fission products which are emitted later do not reach distant points until after the shock wave has passed. Following the shock wave there is a period when the air density is low; gamma-rays travelling through it are much less attenuated and therefore greater gamma-ray intensities are reached at longer distances. Had these gamma-rays travelled through air of normal density, a weapon of much higher explosive yield would have been required to produce the same dose at a given distance. For this reason the hydrodynamic enhancement is usually expressed in terms of an increase in the effective yield of the weapon. Table 11 shows the calculated 'effective' yields of thermonuclear weapons for three actual yields at different distances from the explosion. As is seen, at large distances the effective yield of a weapon can be several thousand times higher than the actual yield, as far as the dose rate from fission product gamma-rays is concerned.

For the reasons stated above only sources 4 and 5 of table 10 need to be considered when calculating gamma-ray doses; the contribution from the other sources is by comparison negligible.

Table 11. Hydrodynamic enhancement of gamma-rays as expressed by 'effective' yield of thermonuclear weapons (low air burst)

Actual yield (Mt)	Distance from explosion (km)	Effective yield (Mt)
0.1	1	0.25
	2	0.35
	3	0.40
1	2	20
	3	30
	4	40
10	3	7 000
	4	20 000
	5	40 000

The attenuation of gamma-rays in passing through the air has the same character as that for neutrons—namely, an exponential fall-off in intensity, superimposed on an inverse square reduction—but the rate of absorption is less than for neutrons. Thus, the exponential absorption of the gamma-rays is such that a tenfold decrease in intensity occurs in 730 metres of air (of density 1.16 kg m^{-3}). Figure 13 gives the gamma-ray dose from sources 4 and 5 as a function of slant distance for a fission bomb of 1-kt yield.

Figure 13. Gamma-ray dose (surface tissue) from 1-kt fission bomb as a function of slant distance

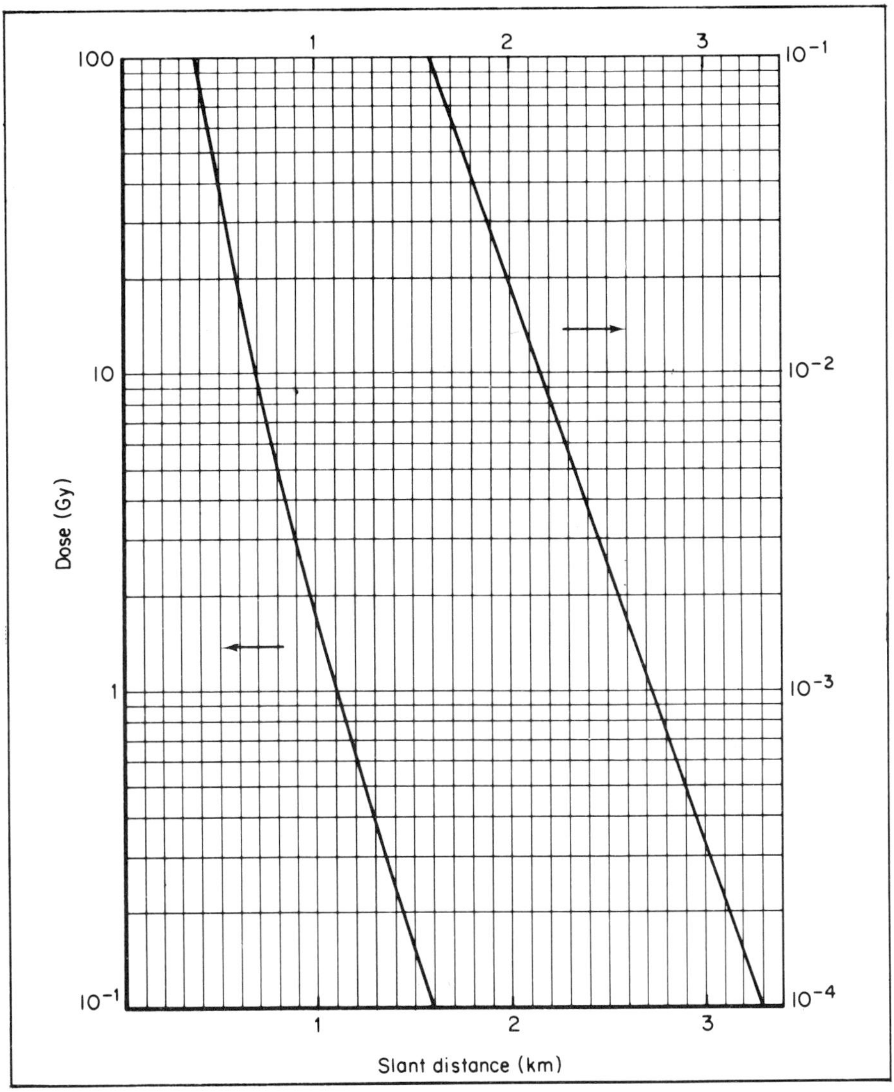

Due to the hydrodynamic enhancement factor, which increases in importance with yield and distance, one cannot use the curves of figure 13 to calculate the gamma-ray doses from weapons of higher yield simply by multiplying by the yield; a correction for the enhancement factor (table 11) would have to be made. Another way is to draw separate curves for each yield. Several such curves are shown on figure 14. Figure 14a applies to fission bombs and figure 14b to thermonuclear weapons with a 50 per cent contribution from fission. The doses in figure 14 are total gamma-ray doses from all sources listed in table 10. For low-yield weapons the dose due to the gamma-

Figure 14. Gamma-ray doses (surface tissue) as a function of slant distance for: (*a*) fission bombs of specified yield, and (*b*) thermonuclear bombs of specified yield

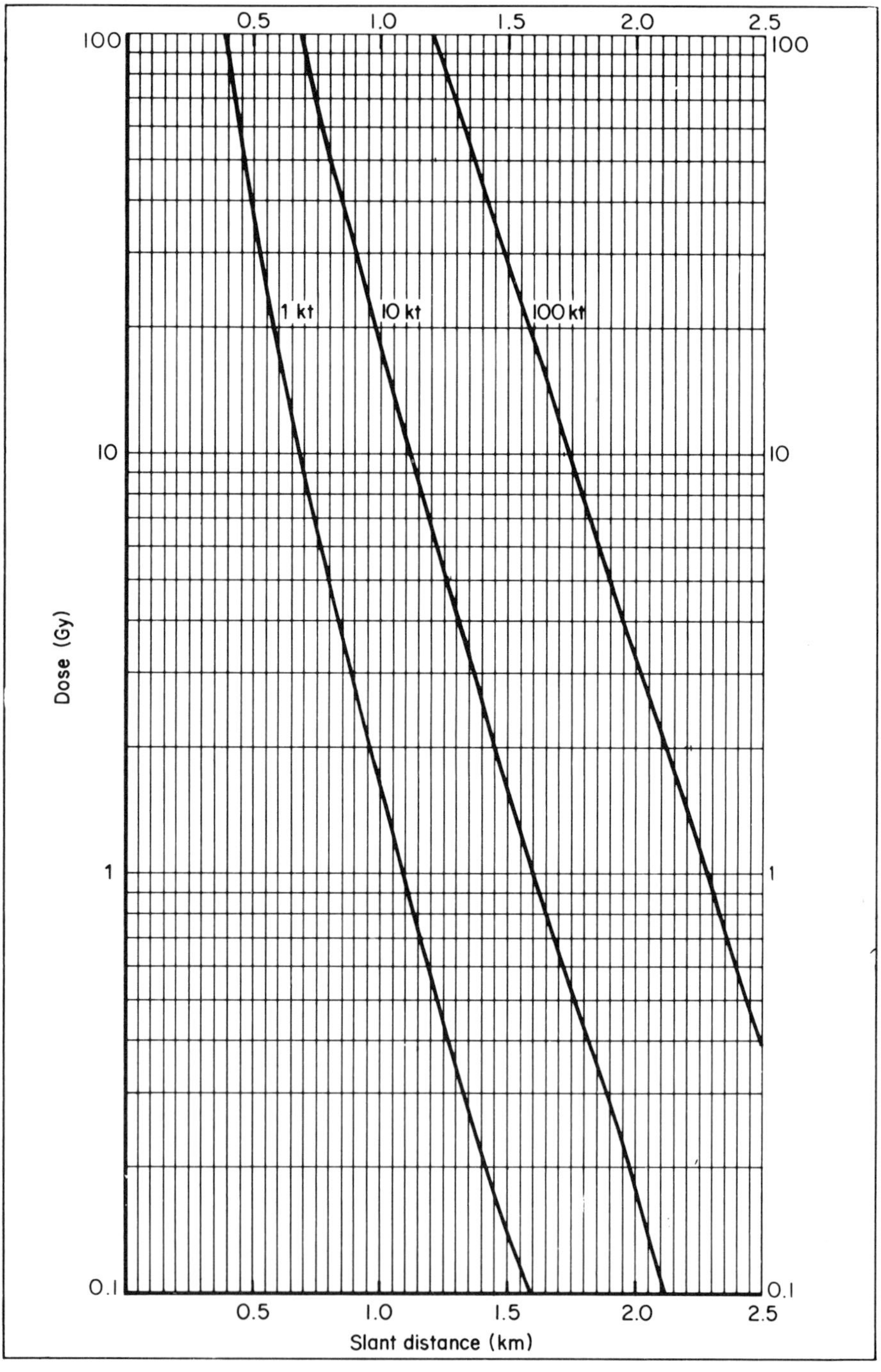

(a)

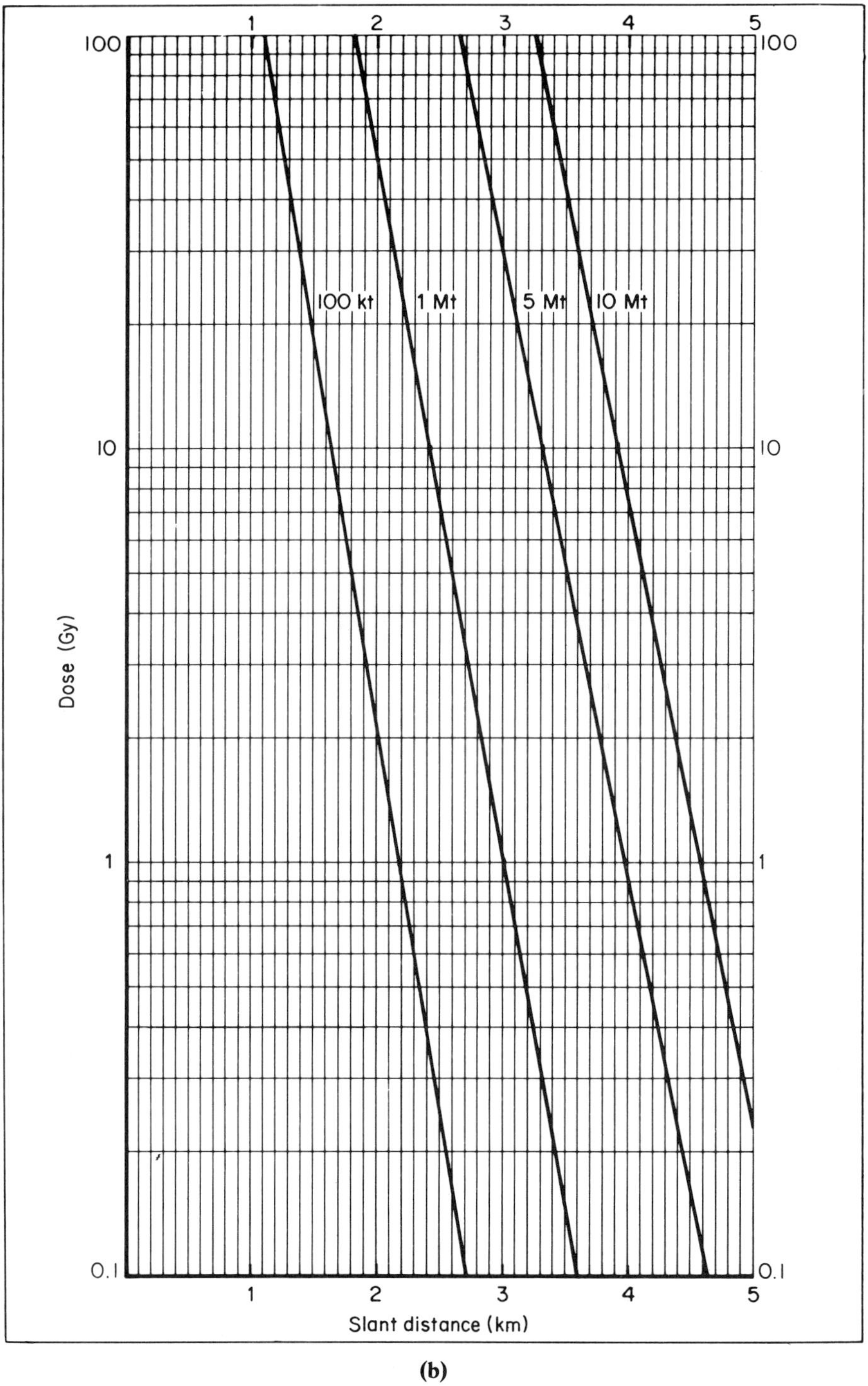

(b)

rays from the fission products is less than that from source 4; for high yields and longer distances the dose is predominantly due to the fission products.

The data in figures 13 and 14 apply to an air burst at a height not less than 100 metres. For a surface burst the doses should be multiplied by a factor which depends on the yield; it ranges from two-thirds for a 1-kt weapon to three for a 20-Mt bomb.

As in the case of neutron doses, it should be emphasized that all estimates of the gamma-ray doses may be wrong by a considerable factor either way. It must also be stressed that the doses given in figures 12, 13 and 14 refer to an exposure received by a person in the open. The reduction in the dose due to the shielding by walls of buildings, or by shelters, is discussed in section 6.4.

4.1.3. Combined doses from neutrons and gamma-rays

Both components of the initial radiation, neutrons and gamma-rays, contribute to the total dose which a person may receive. Owing to their different rates of attenuation in air the relative contributions to the tissue dose, measured at the surface of the body, vary with distance from the explosion. At short distances the neutron dose is greater than that from the gamma-rays, but with increasing distance the gamma-ray dose progressively becomes the dominant contributor. For weapons of high yield, and at long distances, when the hydrodynamic enhancement factor is large, the dose is almost entirely due to gamma-rays.

Because of the greater biological effectiveness of neutrons for long-term radiation effects the dose equivalent (in sieverts) should be used, that is, the neutron dose must be multiplied by the quality factor and the product added to the gamma-ray dose. But for acute effects the quality factor for neutrons can be taken as unity, and the total dose is obtained by simply adding the neutron and gamma-ray doses (in grays). From figures 12 and 14 the distance at which a given total dose will be received from a weapon of given yield can be calculated. Figure 15 shows such a curve for a total tissue-surface dose of 4.5 Gy; this is the LD-50 value for a healthy adult. Under war conditions it is likely that the majority of persons exposed to such a dose would die from the acute effects of radiation. In this sense the curve on figure 15 can be considered as defining the lethal distance for acute radiation effects.

A comparison of this curve with those of figures 3 and 4 (pages 15 and 17), which give the lethal distances for thermal and blast effects, shows that for radiation the lethal distances increase much more slowly with explosive yield. The reason for this is the exponential character of the absorption of neutrons and gamma-rays, which causes a much faster reduction with distance of the radiation effects than for thermal or mechanical damage. As a result of this, for large bombs, when the heat and blast effects reach long distances, the injurious action of neutrons and gamma-rays occurs in an area already devastated by the heat or blast waves. For small bombs the reverse holds: the lethal action of radiation covers a larger area than that affected by the other agents. This can be seen in table 12, which gives the lethal areas (as defined in section 2.2.3) for blast, heat and radiation effects for bombs of various yields when exploded at

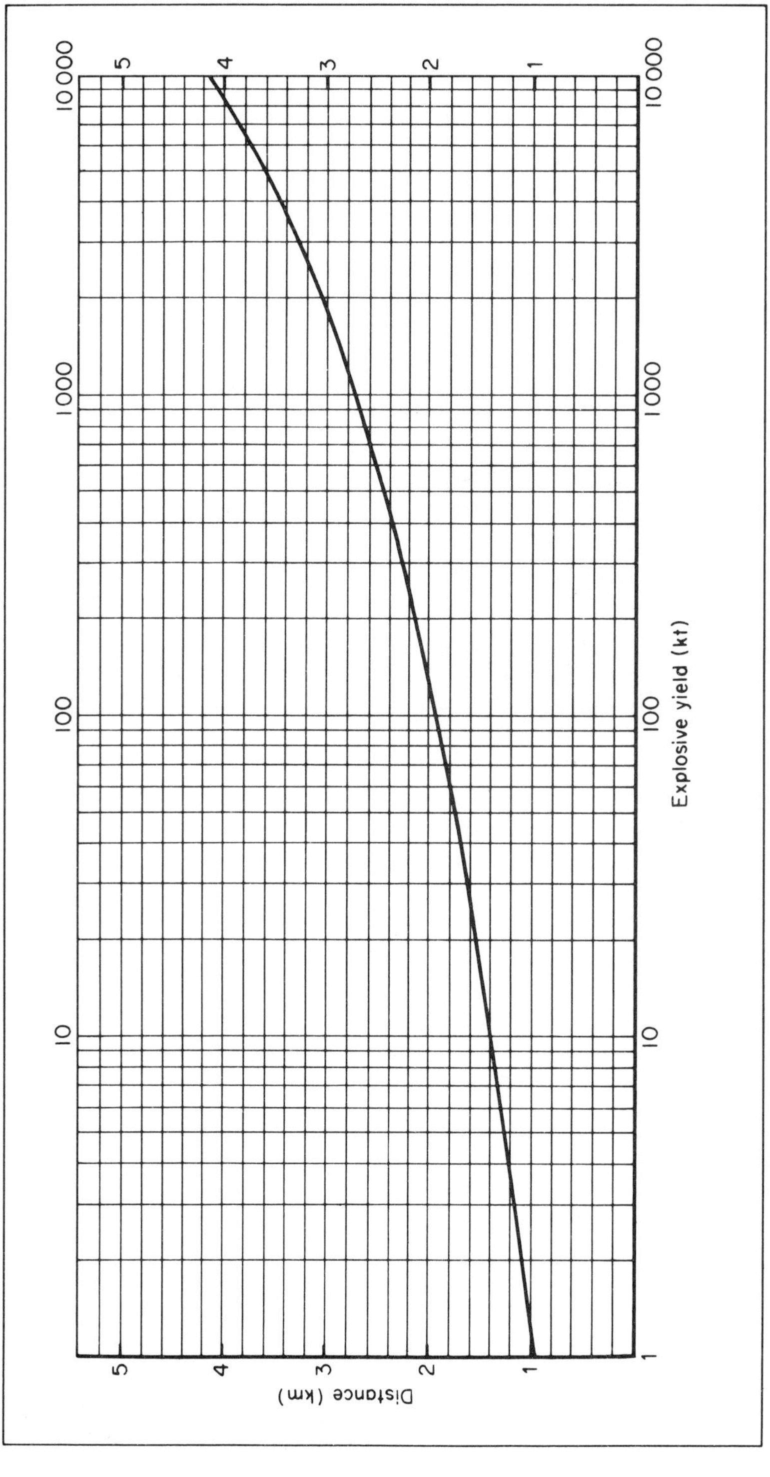

Figure 15. Distances for lethal radiation doses (neutrons plus gamma-rays)

low heights above ground. It is seen that for a 10-kt bomb the lethal areas for blast and radiation effects are nearly the same.

Table 12. Areas of lethal damage from various effects (in km^2)

Type of damage	Explosive yield 1 kt	10 kt	100 kt	1 Mt	10 Mt
Blast	1.5	4.9	17.7	71	313
Heat	1.3	11.2	74.2	391	1 583
Radiation	2.9	5.7	11.5	22	54

4.1.4. The neutron bomb

From the point of view of the lethal effects of the initial radiation the enhanced radiation warhead (ERW), popularly known as the neutron bomb, is in a special category. The specific purpose of the neutron bomb is to kill by radiation but with very little collateral damage. Its explosive yield is about 1 kt and it is intended to be employed with the Lance missile (one of the family of tactical nuclear weapons) and 203-mm artillery shells.

The neutron bomb is a thermonuclear device without a tamper and with a minimum of extraneous material, so that few neutrons will be lost in collisions with nuclei before being emitted. At a given distance from the explosion the neutron fluence is therefore considerably greater than from a fission or F-F-F bomb of the same explosive yield; the energy spectrum also contains a much larger proportion of high-energy neutrons.

In theory the neutrons could carry 80 per cent of the total energy released (see section 2.1.6), but in practice the proportion will be smaller, perhaps 50–60 per cent, because a fission trigger is still needed to initiate the thermonuclear reaction. The energy distribution in an ERW may be as follows: 40 per cent blast, 25 per cent heat, 30 per cent initial radiation and 5 per cent residual radiation. A comparison with the energy distribution in other types of nuclear weapon (made in section 2.1.7) shows that the proportion of energy carried by the neutrons may be as much as 10 times greater in the neutron bomb.

Figure 16 gives the neutron dose as a function of slant distance for a 1-kt ERW. A comparison with figure 12 shows that at short distances the dose from a 1-kt neutron bomb is about the same as from a 10-kt fission bomb. At greater distances the ERW dose becomes progressively higher than that from a 10-kt fission bomb; this follows from the smaller attenuation in air of the higher energy neutrons in the ERW.

If a neutron bomb is exploded at a height of about 500 metres the damage from the blast wave will be quite small; the thermal effect will also be small. But the radiation level will still be high enough to deliver a neutron dose between 100 and 1 000 Gy; such a dose would cause immediate incapacitation and death within a short time. The shielding of the walls of an ordinary tank might reduce the dose by a factor of 2, not enough to prevent the death of the crew. This is indeed the main purpose of the neutron bomb, to produce a minimal amount of material damage but a lethal dose to the crews of tanks.

Figure 16. Neutron dose from a 1-kt neutron bomb

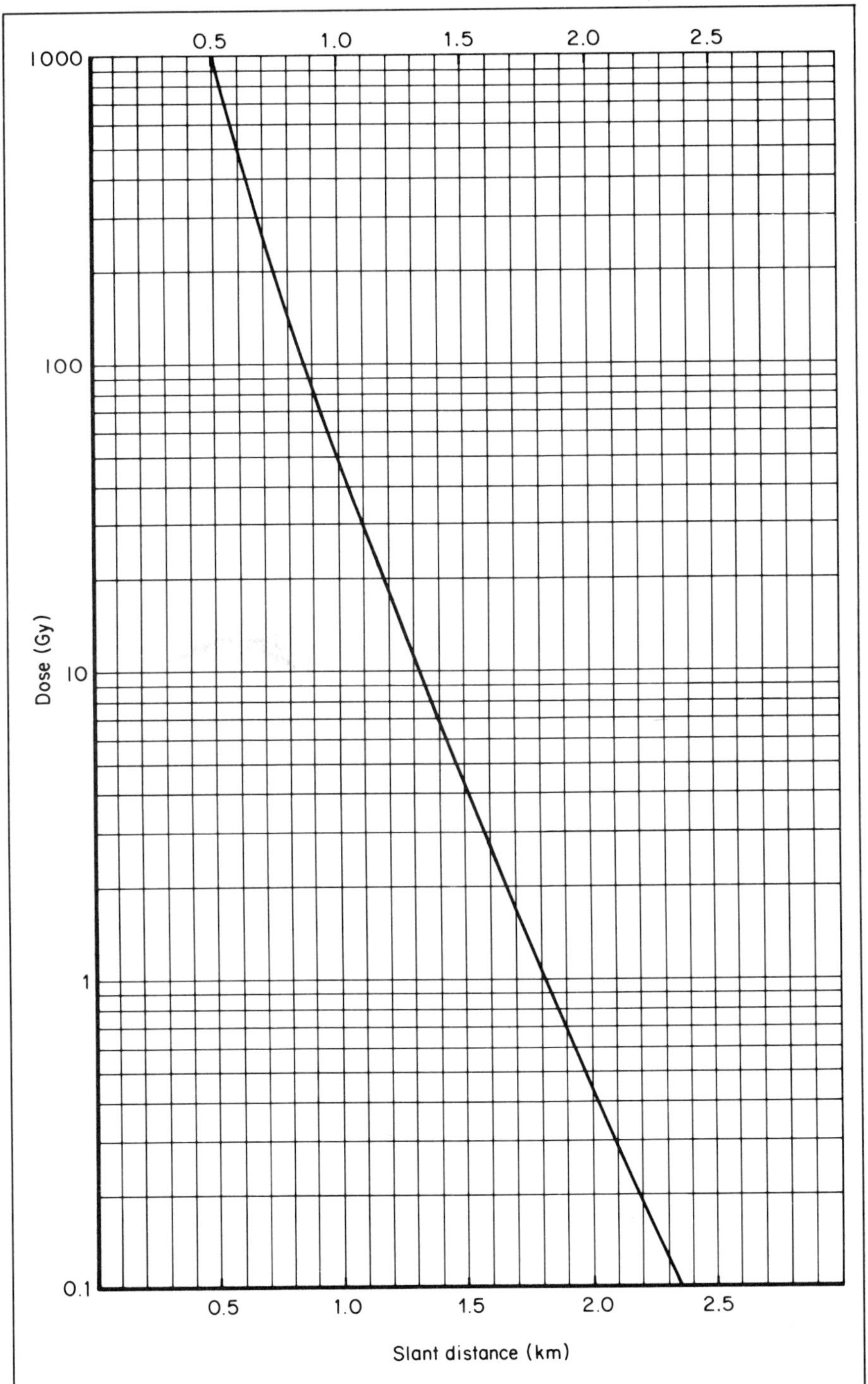

It is this aspect, the killing of people and saving of property, that caused a wave of revulsion and protest when the proposal to deploy the neutron bomb was first announced. It is difficult to see the logic of such protests. If the neutron bomb were surrounded by a uranium tamper it would become an ordinary F-F-F bomb, one of a large number of such weapons already in the arsenals and apparently accepted by society; yet it would then have the potential to kill a much larger number of people by fall-out.

A valid criticism of the neutron bomb is related not to the ethical aspect but to its political and military aspects. The absence of material destruction may encourage the use of the neutron bomb on the territory of an ally, in conditions under which the deployment of ordinary nuclear bombs would have been vigorously opposed. The threshold for the use of nuclear weapons would thus be considerably lowered, with the danger of the conflict escalating into a full-scale nuclear war.

With regard to the military aspect, it may transpire that the level of collateral damage resulting from the employment of neutron bombs, both in human lives and property, will be greater than anticipated. Since the adversary in the conflict will know that his tanks may be attacked with neutron bombs he is likely to take measures to reduce their effectiveness. The tanks may be provided with neutron-absorbing materials to reduce the dose to the crew to levels below immediate incapacitation, even though death will still occur later. The tanks may also be deployed in a formation with much wider spacing. The response to both these measures is likely to be the deployment of a much larger number of neutron bombs; indeed, barrages of many hundreds of neutron bombs are being envisaged. But the result of this would be that a large number of civilians would receive lethal or sub-lethal doses, particularly if the tanks were deliberately deployed near built-up areas; some European countries have an almost continuous urban character. Should some of the neutron bombs accidentally explode near ground level, there would also be damage from blast and heat effects.

As is seen in figure 16, at a distance of 1.5 km the dose in the open would be about 4.5 Gy—an LD-50 dose. If, therefore, an array of neutron bombs were used, even if spaced at intervals as large as 2.3 km, the radiation level in the whole area covered by the array would be sufficient to give a lethal dose to all persons caught in the open.

Another presumed military advantage of the neutron bomb—that owing to the absence of fall-out the troops of the side that used the ERW would be able to enter the battlefield immediately afterwards—would also be lost if a large number of bombs were used, because of the intense radioactivity induced by the neutrons in the soil and other materials.

From the point of view of the initial radiation each such neutron bomb would be nearly the same as the Hiroshima bomb, and as experience there has shown, persons entering the area after the explosion can receive a dose high enough to cause radiation sickness, quite apart from long-term effects.

Finally, if the side at the receiving end of the neutron bomb did not possess it in its arsenal, the obvious retaliation would be by the use of ordinary nuclear weapons; there was in fact a warning by the Soviet Union that any use of nuclear weapons would bring a massive retaliation with nuclear weapons. The door to a widespread nuclear war would thus be opened.

Summing up, the big military advantages claimed for the neutron bomb appear to be largely spurious. The main result of its deployment would be a lowering of the nuclear threshold and an increased probability of an all-out nuclear war.

4.2. Fall-out

Ionizing radiation emitted later than one minute after the explosion is called residual radiation; most of it comes into the category of fall-out. The radioactive products of fission may be deposited on the ground at distances well beyond the range of destruction and casualties caused by the blast, heat and initial radiation. They may reach these regions and present a radiation hazard long after the explosion. With high-yield bombs a single detonation may contaminate a huge area with radioactivity and make it uninhabitable for a long time. Fall-out is therefore a major and unique effect of nuclear weapons.

It should be noted that *all* nuclear weapons detonated in the air give rise to fall-out, but where and when it occurs depends primarily on the altitude of the explosion.

With explosions in the air at such high altitudes that the fire-ball does not touch the ground (see section 2.2.1 and figure 2, page 13), the fission products, which are initially in gaseous form, rise with the fire-ball to great heights into the stratosphere. When the temperature of the fire-ball becomes sufficiently low the radioactive materials form particles through condensation and coagulation. These particles are very small and because of this their descent is very slow; it may take many months before they come down to the ground. By that time they have been carried by stratospheric winds round the globe. Under such conditions there is no fall-out, or very little, in the vicinity of the explosion. This type of fall-out is referred to as 'global' or 'delayed' fall-out.

With surface explosions, or at altitudes low enough for the fire-ball to touch the ground, huge quantities of earth and debris are sucked up with the fire-ball together with the fission products. As the fire-ball cools the radioactivity condenses on the particles of the ground material; many of these are large particles and they come down by the force of gravity within a day or so, at distances not too far from the burst (some hundreds of kilometres). This constitutes the 'local' or 'early' fall-out and it contains more than half of the total radioactivity. The extent and location of the early fall-out depend primarily on the meteorological conditions, mainly on the speed and direction of the wind. They also depend on precipitation conditions; should the airborne debris be intercepted by a rain cloud some of the particles may come down to Earth with the rain or snow; this is called 'rain-out' or 'snow-out'.

These different types of fall-out create radiation hazards to man of entirely different magnitudes: a large hazard to a relatively small number of people in local fall-out, and a much smaller hazard but to a much larger number of people in remote countries from the global fall-out. These differences in the magnitude of the hazard result largely from the physical properties of the fall-out material.

4.2.1. Nature and quantity of radioactivity in the residual radiation

Most of the residual radiation is in the form of fall-out originating from the radioactivity of the fission products. However, the fall-out also contains radioactivity induced by neutrons in the weapon materials and in the air, as well as the uranium and plutonium which did not undergo fission and the tritium from fusion reactions. In addition to the fall-out there is some residual radiation in the vicinity of the explosion from the radioactivity induced by the neutrons in the ground and in walls of buildings.

4.2.1.1. Fission products

When a uranium or plutonium nucleus undergoes fission it splits into two parts. (Rarely it can also split into three parts; this is called ternary fission.) These fragments are nuclei of lighter elements such that the sum of their atomic numbers is equal to that of uranium (92) or plutonium (94). For example, uranium may split into nuclei of the elements krypton and barium which have atomic numbers 36 and 56; similarly, plutonium may split into selenium and neodymium (atomic numbers 34 and 60). Uranium or plutonium can split in some 40 different ways, giving rise to about 80 different nuclides in the first instance. However, not all ways of splitting are equally probable; some occur much more frequently than others. Thus, there is a distribution of the frequency with which the initial fission products occur. This distribution varies with the type of nucleus undergoing fission (^{235}U, ^{238}U or ^{239}Pu) and with the energy of the neutrons which gave rise to the fission. Figure 17 shows the distribution of the fission products which occurs when ^{238}U is bombarded with fission neutrons, that is, neutrons with a continuous spectrum of energy (as shown in figure 11, page 60). The ordinates of figure 17 give the fission yield, that is, the percentage of all fissions which result in products of the given mass number. (The mass number is the relevant parameter here, since in a chain of radioactive transformations occurring by beta-decay the mass number of all nuclides in the chain remains the same.) It is seen that the yield for some modes of fission is nearly 7 per cent, while for other modes it is less than 0.1 per cent. It is of interest to note that symmetric fission, that is, a split into two identical nuclides, occurs very rarely.

Most of the fission fragments are radioactive and undergo decay with a half-life which is a characteristic property of the given nuclide. In most cases the product of the decay is also a radioactive nuclide, which in turn undergoes disintegration, producing a nuclide which may again be radioactive; this goes on until a stable nucleus is formed. On the average a chain of four radioactive nuclides is formed for each primary fission fragment. Table 13 gives two

Figure 17. Fission yields for products of given mass number. (Fission in ^{238}U by fission neutrons.)

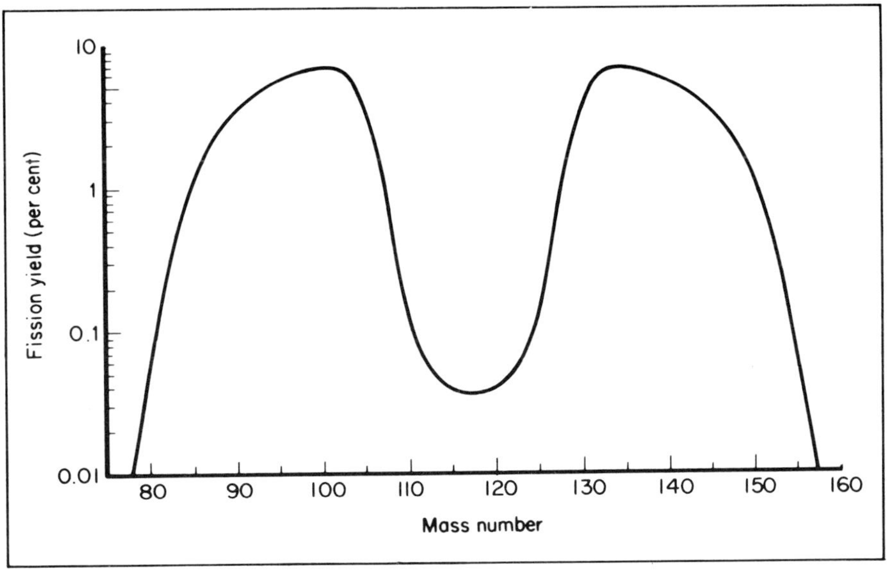

examples of such chains, one with five and the other with three radioactive nuclides; their half-lives are shown in parentheses. Altogether some 300 different radioactive nuclides, involving 36 elements (from zinc to terbium), are formed in fission. Their half-lives vary from a fraction of a second to many millions of years.

Table 13. Radioactive chains of fission products[a]

^{90}Br(1.9 s) → ^{90}Kr(32 s) → ^{90}Rb(2.6 m) → ^{90}Sr(28.8 y) → ^{90}Y(64 h) → ^{90}Zr (stable)
^{137}I(24 s) → ^{137}Xe(3.8 m) → ^{137}Cs(30.2 y) → ^{137}Ba (stable)

[a] Half-lives: s–seconds; m–minutes; h–hours; y–years.

In practically all cases the disintegration results in the emission of beta-rays (fast electrons), often accompanied by gamma-rays. From the point of view of radiation hazards some of the fission products are more important than others; the relevant parameters are the fission yield, the half-life and the type and energy of the radiations emitted. Table 14 gives a list of the more important fission products together with some of their characteristics.

The intensity of a radioactive substance is measured in terms of the number of nuclides which undergo disintegration per unit of time and is called the activity. The unit of activity is the becquerel (Bq), which is equal to the number of disintegrations occurring per second. The old unit, the curie (Ci), initially based on the activity of 1 g of radium-226, is equal to 3.7×10^{10} disintegrations per second. Therefore, 1 Bq = 2.7×10^{-11} Ci.

The activity of a substance is proportional to the total number of atoms in the substance and therefore to its mass. But it also depends on the half-life: the

Table 14. Important fission products

Nuclide	Half-life	Type of radiation		Percentage fission yield
Krypton-85	10.7 years	β	γ	0.2
Strontium-89	50.5 days	β	–	2.9
Strontium-90	28.8 years	β	–	3.2
Yttrium-91	58.5 days	β	γ	5.8
Zirconium-93	1.5×10^6 years	β	–	6.4
Zirconium-95	64.0 days	β	γ	6.3
Technetium-99	2.1×10^5 years	β	–	6.3
Ruthenium-103	39.4 days	β	γ	6.6
Ruthenium-106	367 days	β	–	2.7
Iodine-129	1.6×10^7 years	β	γ	0.9
Iodine-131	8.04 days	β	γ	3.2
Xenon-133	5.25 days	β	γ	5.5
Caesium-134	2.06 years	β	γ	6.6
Caesium-135	3×10^6 years	β	–	6.0
Caesium-137	30.2 years	β	γ	6.2
Barium-140	12.8 days	β	γ	5.7
Cerium-144	284 days	β	γ	4.9
Promethium-147	2.6 years	β	γ	2.4
Samarium-151	90 years	β	γ	0.5

longer the half-life the smaller the proportion of nuclides which break up per unit time. Taking these two factors into account it is simple to calculate the activity of a given mass of radioactive substance. The activity of 1 g of any radioactive material is obtained by dividing 4.23×10^{23} by the mass number of the nuclide and by its half-life in seconds. For example, the activity of 1 g of iodine-131 (half-life 8.04 days = 7×10^5 seconds) equals

$$\frac{4.2 \times 10^{23}}{131 \times 7 \times 10^5} = 4.6 \times 10^{15}\ \text{Bq}\ (\sim\ 124\ \text{kCi}).$$

Similarly, the activity of 1 g of plutonium-239 (half-life 24 100 years = 7.6×10^{11} seconds) comes out to be

$$\frac{4.2 \times 10^{23}}{239 \times 7.6 \times 10^{11}} = 2.3 \times 10^9\ \text{Bq}\ (\sim\ 62\ \text{mCi}).$$

Using this formula one can calculate the initial activity of each of the fission products listed in table 14, resulting from the explosion of a fission bomb of given yield. For example, recollecting that in the detonation of a 1-kt bomb 57 g of fissile material is produced (section 2.1.1), it is calculated that the initial activity of caesium-137 is 6.7×10^{12} Bq per kiloton.

The main hazard from fall-out—at least in the first few weeks—is due to the gamma-rays emitted from the fission products. For each kiloton of explosive yield of a fission bomb, the total activity in terms of gamma-ray emission is about 2×10^{19} Bq at one hour after the explosion. This activity decreases rapidly with time; the rate of this decrease is discussed in section 4.2.2.

4.2.1.2. Induced radioactivity in fall-out

Apart from the fission products the fall-out also contains radioactive substances produced by the interaction of neutrons with the materials of the bomb assembly and with nitrogen of the air. In the latter case the interaction results in the production of carbon-14. This nuclide has a long half-life (5 730 years) but emits only very soft beta-rays and no gamma-rays; it therefore presents only an internal hazard.

Carbon-14 occurs naturally in very small quantities, being produced by the interaction of the neutrons from cosmic radiation with nitrogen in the air. The rate of natural production is about 10^{14} Bq per year. In nuclear explosions the production rate is about 1.5×10^{12} Bq per kiloton. As a result of the atmospheric tests of bombs up to 1963 the amount of carbon-14 has greatly increased and the atmospheric inventory has nearly doubled. Owing to the uptake of carbon-14 by the oceans and the biosphere the amount of it in the atmosphere is decreasing, but it will continue to contribute to the natural background of radiation for many years, even if no further amounts are injected from nuclear weapons (see section 4.2.6).

Of the radioactivity induced in the materials of the bomb assembly the most important nuclide is iron-55. It has a half-life of 2.7 years and decays by electron capture with the emission of X-rays. The rate of production of iron-55 is of the order of 10^{13} Bq per kiloton. Because of the concentration of iron in red blood cells, iron-55 adds significantly to the radiation hazard. It also concentrates in ocean fish and the hazard is therefore greater for people consuming large quantities of fish.

A potentially dangerous radioactive nuclide, which might be induced by neutrons from the bomb, is cobalt-60. It has a half-life of 5.3 years and emits gamma-rays of high energy (0.4 pJ per disintegration). If a large amount of ordinary cobalt were incorporated into the assembly of the bomb, a huge activity of cobalt-60 would become part of the fall-out and it would produce hazardous radiation levels for a much longer time than the fission products. It is not known whether this potential of cobalt has been tested in weapons, but at one time the possibility of the cobalt bomb as a doomsday machine was much discussed [13].

4.2.1.3. Uranium and plutonium

In fission bombs some of the uranium-235 or plutonium-239 is left unfissioned when the device explodes; these fissile materials are then dispersed together with the fission products. In thermonuclear weapons with a ^{238}U tamper the capture of neutrons by the uranium results in the formation of other nuclides, which are also dispersed, and which are subsequently transformed into plutonium. The amount of plutonium produced in this way depends on the energy of the neutrons. The primary neutrons from the fusion reactions have a high energy and for these the probability of capture is small. But after the neutrons have made a number of collisions with the hydrogenous material of the bomb they are slowed down sufficiently, before the assembly breaks up, to produce large quantities of plutonium. The overall probability of neutrons causing fission in the ^{238}U tamper is about the same as for their capture to produce plutonium, and calculations show that for a 10-Mt bomb, with half of the yield due to fission, the amount of plutonium produced would be about 250 kg; this is much more than the plutonium that was used for the trigger of the bomb.

Owing to the very long half-life of plutonium-239 (24 100 years) the activity is very small. The above quantity would have an activity of about 6×10^{14} Bq, or 10^{11} Bq per kiloton of fission yield. A comparison with the activities of fission products (see section 4.2.1.1) shows it to be a very small fraction of the initial activity of the fission products. On the other hand, the very long half-life means that the plutonium will remain active long after almost all of the fission products have decayed.

Despite its low activity, plutonium-239—as an alpha emitter—presents a significant internal radiation hazard. Most of the hazard would result from the inhalation of air containing insoluble plutonium particles. Ingestion of food contaminated with plutonium would add to the hazard. As was shown in section 3.7, even very small amounts of plutonium in the lungs can produce severe acute, as well as long-term, radiation effects.

Large quantities of uranium-238 from the tamper would also be dispersed with the fall-out, but owing to the extremely long half-life of this nuclide (4.5×10^9 years) the activity would be very small and its radiation hazard negligible.

4.2.1.4. Tritium

Tritium is an essential ingredient of thermonuclear weapons, and what is left unused in the reactions is released after the explosion of a thermonuclear bomb. A small proportion of fissions also results in the production of tritium. In addition, some tritium is produced by the interaction of high-energy neutrons with the nitrogen in the air.

Tritium has a half-life of 12.3 years. A small amount of it occurs naturally in equilibrium with the stable isotopes of hydrogen, as it is being continuously produced by the action of cosmic ray neutrons with deuterium. But the amount of tritium released in nuclear weapons far exceeds the natural occurrence. Thus, following the period of heavy testing of nuclear weapons until 1963 the amount of tritium released into the atmosphere was about 100 times greater than the annual natural production. Since then the amount of tritium has decreased by natural decay but considerable quantities of tritium from the tests are still present in the stratosphere, from where it is gradually descending (see section 4.2.6).

Tritium released from weapons is rapidly converted into tritiated water; this differs from ordinary water in that an atom of tritium replaces an atom of ordinary hydrogen in the water molecule (HTO instead of H_2O). Tritiated water is about 11 per cent heavier than ordinary water but it follows the latter in ecological and physiological processes. As a result, it is incorporated into the human body through food and drink and thus presents an internal radiation hazard. As an external hazard it is negligible, since tritium emits only beta-rays of very low energy and no gamma-rays. Even the internal hazard is small compared with that from some fission products (e.g., strontium and caesium) because of the large dilution of tritiated water with ordinary water and also because of its relatively short half-life.

4.2.1.5. Induced radioactivity near ground zero

Apart from the radioactivity which is carried by the winds and later deposited on the ground as fall-out, there is also radioactivity induced in the immediate vicinity of the explosion. The neutrons from the initial radiation, on hitting the ground or walls of buildings, render these materials radioactive. Some of these radioactive nuclides emit gamma-rays. Thus, even when the explosion occurs at such a height that there is no fall-out in the vicinity of the explosion, this induced radioactivity presents a radiation hazard for some time afterwards. In the Japanese cities, where owing to the height of the explosions (over 500 metres) there was no fall-out (except for a rain-out in some localities), the persons who entered the cities during the first three days received significant doses of radiation, and some late effects, such as leukaemia, were reported to have been induced by this residual radiation [14].

The important radioactive substances induced by neutrons are aluminium-28, manganese-56 and sodium-24, which are common in soils and in building materials. Should the neutrons hit sea water, the activity of sodium-24 and chlorine-38 would be particularly intense. All of these nuclides emit gamma-rays. However, they all have short half-lives, ranging from about 2 minutes to 15 hours. Thus, even with an initially intense induced radioactivity, after one week the activity will be reduced to a negligible value.

The amount of induced radioactivity depends on the number of neutrons reaching the ground. With high-altitude bursts, when there is no local fall-out, the neutrons will be greatly attenuated before reaching the ground and the induced radioactivity will be correspondingly small. With near surface bursts, when the number of neutrons hitting the ground is large, there will be intense induced radioactivity, but under these conditions there will also be a strong local fall-out and the activity from the fission products will be much higher than that induced by neutrons. The situation is different with enhanced radiation weapons, which contain relatively little fission but where the neutron intensity is very high near the ground.

But even with fission or F-F-F bombs the induced activity could add significantly to the hazard. Its main effect would be in preventing early rescue work, as persons entering the devastated area soon after the explosion would be exposed to dangerously high levels of radiation. Thus, even when the detonation did not produce local fall-out, radiation could indirectly contribute to the casualty toll. It could also contribute directly: for the persons who received a high but not lethal dose from the initial radiations the additional dose from the induced radioactivity might greatly reduce the chances of survival.

4.2.1.6. Dose rates from fall-out radioactivity

The radioactivity in the fall-out can expose populations in several ways, in different time sequences: (*a*) external irradiation by the radioactive cloud as it passes overhead; (*b*) internal irradiation through the inhalation of radioactive air; (*c*) external irradiation by the gamma-rays from the radioactive substances deposited on the ground; (*d*) internal irradiation through eating meat or drinking milk from animals which had ingested radioactive substances, or by eating vegetarian food from plants which had incorporated such substances, or by drinking contaminated water.

In the case of local fall-out pathway (*c*), that is, gamma-ray exposure from matter deposited on the ground, represents the most important hazard. It gives rise to total-body exposure and the dose rate is proportional to the deposited activity. It also depends on the energy of the gamma-rays: the higher the energy the greater the dose rate. The gamma-rays from the fission products of a bomb have a wide range of energies; moreover, the energy spectrum varies with time after the explosion. At one hour after the explosion an average value of the energy can be taken as 0.1 pJ. If one assumes a uniform deposition of the radioactivity over a large, smooth area, one can calculate the dose rate at a given time and at a given height above ground, say 0.9 metres, which is approximately the mid-height of an adult. Such calculations have shown that if the gamma-radioactivity resulting from the detonation of a 1-kt fission bomb—that is, all fission products plus the induced radioactivity—were deposited on an area of 1 km^2, the dose rate would be 75 Gy per hour at one hour after the explosion.

In this calculation it is assumed that the terrain is flat; any unevenness would reduce the dose rate, since some of the gamma-rays would be then absorbed in the ground. To allow for this effect the dose rate should be multiplied by 0.7.

It should also be noted that the above calculation was made in relation to the total radioactive content of the fall-out. As mentioned earlier, only part of it, about 60 per cent, is deposited as local fall-out.

4.2.2. Decay of early fall-out

The decay of each of the radioactive nuclides among the fission products follows a simple exponential decrease with time (in equal intervals of time, equal proportions of the activity decay), but for the whole mixture of the fission nuclides the variation of activity with time is highly complex. This is due to the fact that the products of the decay of the fission fragments are themselves often radioactive, and the activity of these products builds up from zero and increases to a maximum before beginning to decrease by radioactive decay.

The result is that during the period of build-up the overall decay is slower than it would have been otherwise. It has been found empirically that during the first six months after the explosion the activity, and therefore the rate at which the radiation dose is delivered, decays approximately according to a power law such that for every sevenfold increase in time the dose rate decreases by a factor of 10. For example, at the end of one week the dose rate is 10 times smaller than it was after one day; after seven weeks it is 100 times smaller than after one day. From six months onwards the rate of decay is faster, but there is no simple empirical formula to present it.

Although the residual radiation differs for different types of bomb, as long as the fall-out is due mainly to fission products the variation of dose rate with time is not significantly different.

Table 15 gives the average dose rates due to the gamma-rays from early fall-out at different times after an explosion. The dose rates in the table are with reference to the dose rate at one hour, at which time it was assumed to be 1 Gy per hour. The same data, but extended to longer periods, are presented in figure 18. If in any actual situation the dose rate at one hour, or at any other time, is known, the dose rates at other times can be obtained simply by proportion.

Table 15. Dose rates from fall-out at various times after an explosion

Time (hours)	Relative dose rate (Gy/h)
1	1
2	0.40
4	0.15
6	0.10
12	5.0×10^{-2}
24	2.4×10^{-2}
36	1.6×10^{-2}
48	1.1×10^{-2}
72	6.2×10^{-3}
100	3.6×10^{-3}
200	1.7×10^{-3}
500	5.0×10^{-4}
1 000	2.3×10^{-4}

For example, if it is known that in a given locality the dose rate at 4 hours after the explosion was 1.2 Gy per hour, the dose rate at 36 hours can be calculated as follows. From table 15 we see that the relative dose rates at 4 and 36 hours are 0.15 and 0.016 respectively. Therefore, the dose rate at 36 hours is $1.2 \times 0.016/0.15 = 0.13$ Gy per hour.

It should be noted that the variation of dose rate with time, as given in table 15 or figure 18, is valid only after the fall-out is complete at a given place, and on condition that no additional material is brought into the location by another explosion, and that none is lost by weathering.

The total dose received by a person is obtained by multiplying the dose rate by the duration of exposure. Since, however, the dose rate rapidly

Figure 18. Dose rate from fall-out as a function of time after the explosion

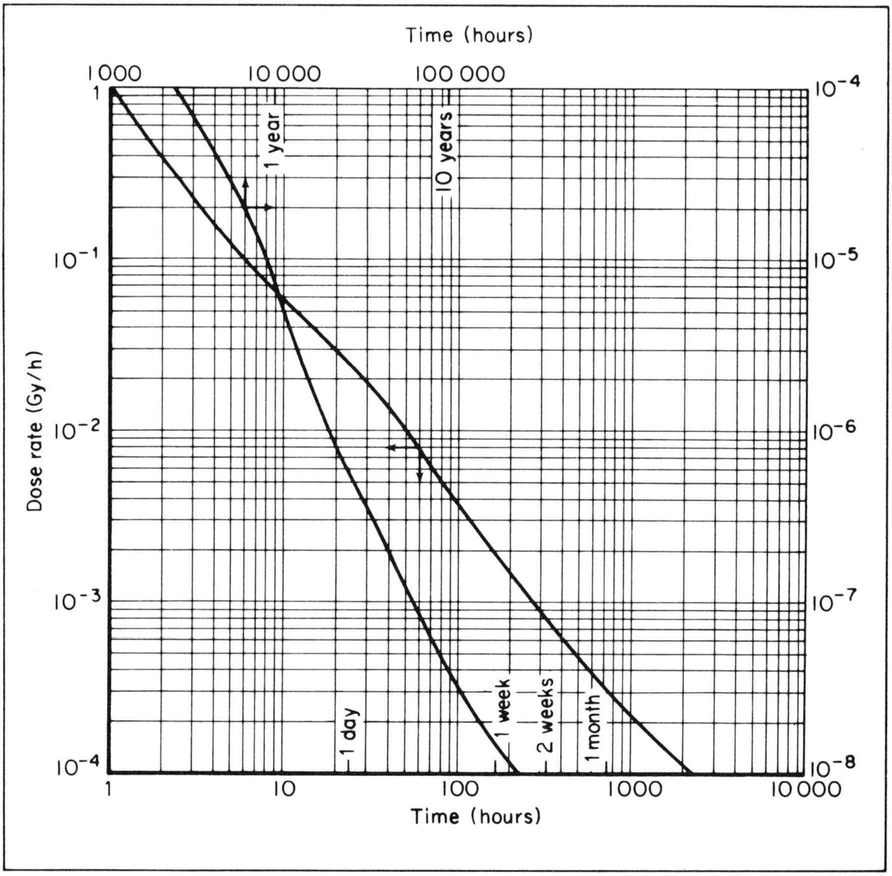

decreases with time, the calculation of the total dose involves an integration over the relevant values of dose rates. The result of such integration is presented in figure 19, which gives the accumulated dose (in grays) starting from one minute after the explosion up to the given time, assuming that the dose rate at one hour is 1 Gy per hour. For other values of the dose rate the values in figure 19 have to be multiplied by the actual dose rate at one hour. For example, suppose that the dose rate measured at 24 hours after the explosion was 0.48 Gy per hour; what was the accumulated dose received by a person during the 24 hours? From table 15 we see that the dose rate at 24 hours is 2.4 per cent of that at 1 hour; therefore, the reference dose rate is $0.48/0.024 = 20$ Gy per hour. Figure 19 shows that the accumulated dose at 24 hours is 7.45 Gy; therefore, the total dose was $20 \times 7.45 = 149$ Gy.

Figure 19 also allows the calculation of the total dose received by a person who enters a given fall-out locality at a certain time after the explosion and remains in it for a certain period. Thus, in the previous example, if a person entered the contaminated area at 12 hours and stayed there for 36 hours the total dose received would be about 16 Gy. This is calculated as follows. The relative accumulated doses at 12 and 48 hours are 7.05 and 7.87 Gy (figure 19); since the dose rate at 1 hour was previously found to be 20 Gy per hour, the total dose is 20 × (7.87 − 7.05) = 16.4 Gy.

Figure 19. Accumulated dose from local fall-out as a function of time after the explosion

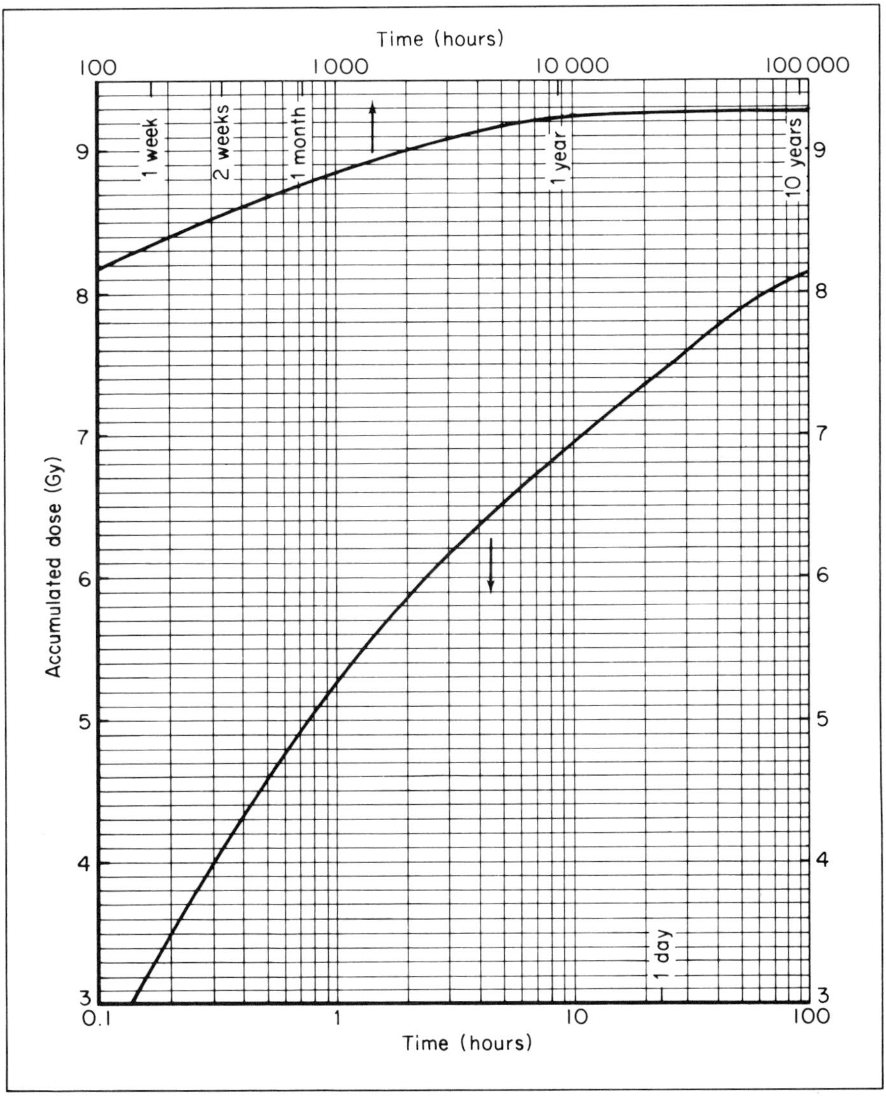

As is seen from figure 19, the total accumulated dose tends towards a finite limit, namely 9.3 Gy if the dose rate at one hour is 1 Gy per hour. This value, 9.3 Gy, gives the infinity dose, that is the total dose accumulated starting from one minute after the explosion until an infinite time.

The infinity dose is used when calculating the total accumulated dose, in a given locality, due to fall-out which was deposited a certain time after the explosion. The following example will illustrate this. In a certain locality, the fall-out was deposited nine hours after the explosion; the dose rate measured at that time in the locality was 0.23 Gy per hour; what will be the maximum accumulated dose in the locality? From figure 18 we see that at nine hours the dose rate is 7 per cent of that at one hour. Therefore, the reference dose rate (at one hour) is $0.23/0.07 = 3.3$ Gy per hour. Figure 19 shows that the accumulated dose, for the reference dose rate, up to nine hours is 6.88 Gy; the remaining dose to be accumulated to infinity is therefore $9.3 - 6.88 = 2.42$ Gy. For the actual dose rate, the dose to be accumulated to infinity is thus $3.3 \times 2.42 = 8.0$ Gy.

This dose was described as the maximum accumulated dose, because it implies that the individual exposed remained in the locality for the whole time in the open. If he spent some time indoors, the accumulated dose would be smaller due to the shielding of the walls.

The method of dose calculation outlined above applies only to a single detonation. If the fall-out in the given locality is caused by several bombs, exploded at different times, the variation of dose rate with time will be quite different from that given in table 15 or figure 18, and a single measurement of the dose rate in that locality would not be sufficient to calculate the accumulated dose.

The rapid rate of decay of the early fall-out is often used to reassure the population that the danger from fall-out radioactivity is over after about two weeks. This is misleading. What matters is not the dose rate but the accumulated dose, and the latter decreases with time less rapidly than the dose rate. Thus, at two weeks, the dose rate is about 1 000 times lower than at one hour (see figure 18), but the infinity dose from two weeks onwards is only 5.5 times smaller than the infinity dose starting from one hour (see figure 19).

In any case, if the initial fall-out in the given area was of high intensity, the exposure level could be dangerously high for a long time and the areas could remain uninhabitable for many years. For example, if the dose rate at one hour was 100 Gy per hour, a person entering the fall-out area after one month and then remaining there could accumulate a dose of about 50 Gy in one year. Even if he entered after one year the dose accumulated during the next year could be about 3 Gy.

4.2.3. Distribution of early fall-out

As mentioned previously, if the fire-ball touches the ground, vaporized earth particles are scooped up with the fire-ball to a great height; the fission products, together with the other radioactive materials, then condense on these particles and begin to descend to the ground. The particles have a variety of sizes, from less than 1 micrometre to over 1 mm in diameter, with a peak in the distribution at about 50 micrometres. The heaviest particles begin to fall back very soon in the immediate vicinity of the burst. The lighter particles stay in the air for a longer period, and may be carried with the wind through very great distances before descending to the ground, the smallest particles travelling furthest. The radioactive materials which are deposited within 24 hours after the explosion constitute the 'early' fall-out. As mentioned above, about 60 per cent of the total radioactivity may be contained in the early fall-out.

For a given bomb the distance travelled by the fall-out particles, and the time and location of their deposition, are primarily dependent on the speed and direction of the wind. As the particles are carried further away by the wind they spread out over larger areas so that the rate at which the radiation dose may be delivered is rapidly reduced with the distance; this decrease is in addition to the radioactive decay which occurs during the time before the fall-out particles reach the ground. Under steady wind conditions, with no shear, at any given time after the explosion the radioactivity will have spread itself in

Figure 20. Contours for fall-out from a 2-Mt bomb at 18 hours after the explosion: (*a*) dose rates, and (*b*) total dose

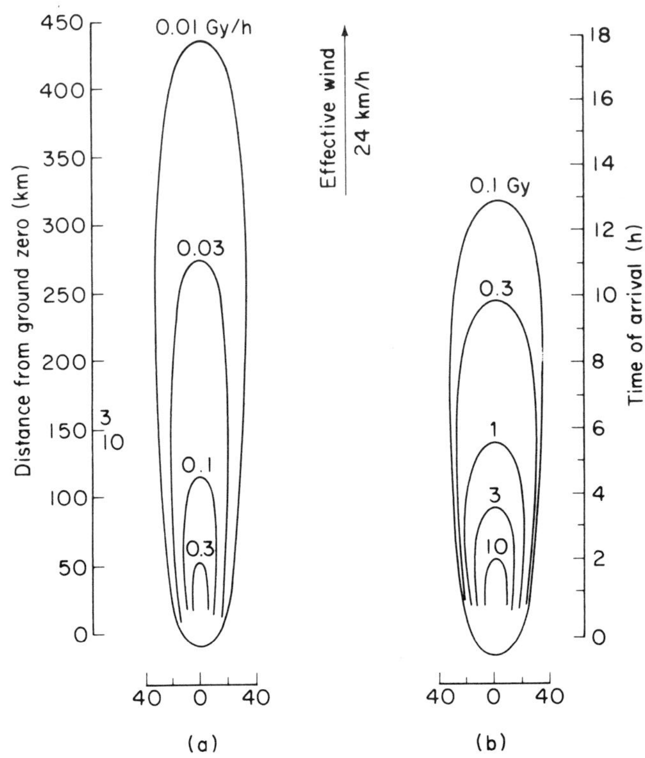

such a way that the lines joining all points with the same dose rate will be cigar-shaped. Figure 20 shows (*a*) several dose-rate contours for the fall-out from a 2-Mt bomb with a 50 per cent fission content at 18 hours after the explosion, when the wind velocity was 24 km per hour; and (*b*) the contours of the total doses which would result from exposure at these dose rates.

It should be noted that these contours represent a transient situation in space and time. At any given place the activity first increases with time, as more fall-out reaches the locality and is deposited, but subsequently it decreases due to natural decay (quite apart from the possible reduction due to weathering). A qualitative idea about these changes can be obtained from figure 21, showing (*a*) the variation of dose rate with time, at two distances, one near and the other remote, downwind from the explosion; and (*b*) the variation of the total dose which a person might receive if he remained for the whole time at these locations.

Figure 21. Dose rates (*a*), and cumulative doses (*b*), (arbitrary units) from early fall-out as a function of time for two distances from the explosion (A – short distance and B – long distance)

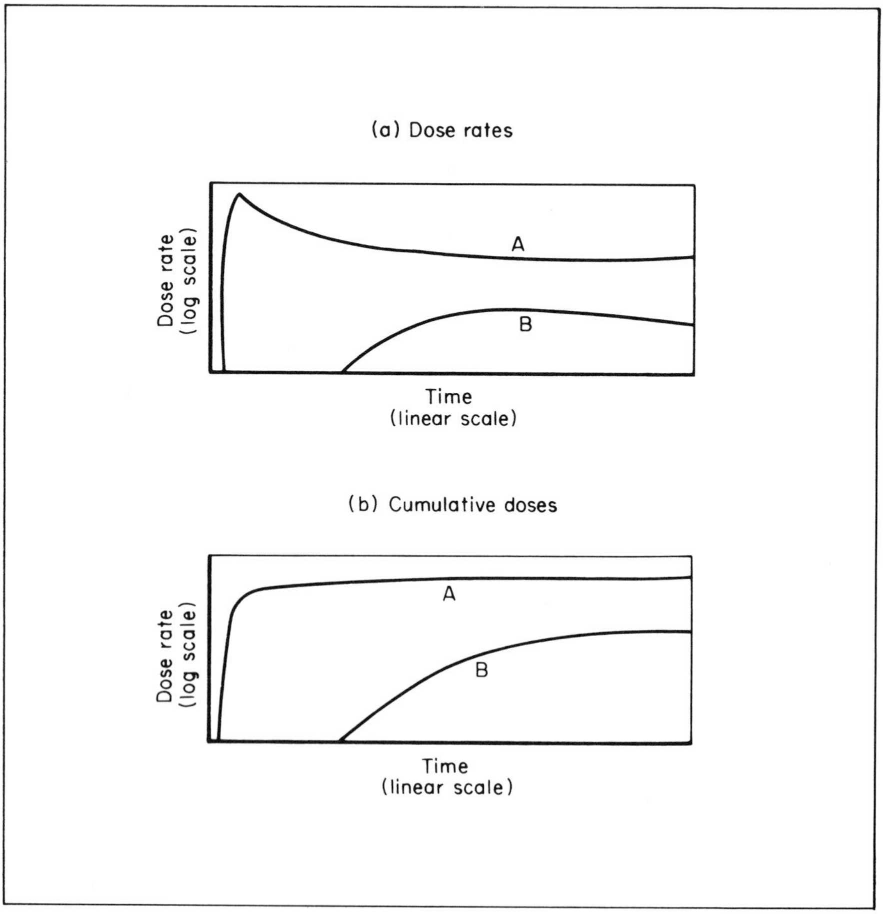

Figures 20 and 21 are intended to illustrate the situation in a grossly simplified manner only. In reality the picture is much more complex because many factors may disturb the pattern of fall-out. First of all, the wind may change in speed or in direction and this would distort the shape of the contours. Then, there may be scavenging of the fall-out by rain or snow if the air-borne particles encounter a region where precipitation is occurring. Such events may give rise to 'hot spots', small areas where a very large amount of radioactivity is deposited; this means, of course, that the fall-out would be depleted in other areas. Rain coming down after the fall-out may wash away some of the activity. All these phenomena have actually been observed in test explosions.

4.2.3.1. Fall-out from a test explosion on Bikini Atoll

An unintended exposure of a population to the radiation from local fall-out occurred in the Marshall Islands, following the Bravo test, 1 March 1954. This was the first test of a large thermonuclear device and the explosive yield was about 15 Mt. The detonation was near the ground, about 2 metres above a coral reef in the Bikini Atoll. After ascending to a height of 35 km, some of the radioactive cloud came down unexpectedly in a long plume in an easterly direction, covering the Marshall Islands, several of which were inhabited by natives and one by United States personnel. An area of about 20 000 km^2 was contaminated to such an extent that lethal doses would have been received by persons staying in the open. Figure 22 shows the outlines of the various atolls, as well as contour lines for several values of total doses which persons might have received during 96 hours after the explosion. (The accumulated doses to infinity would be nearly double those values.) The bottom and top scales give the distance from ground zero and the time of arrival of the fall-out at the given location. Owing to an insufficient number of monitoring instruments the contours were drawn largely by guesswork. In particular, it is possible that the

Figure 22. Dose contours after the Bravo test explosion

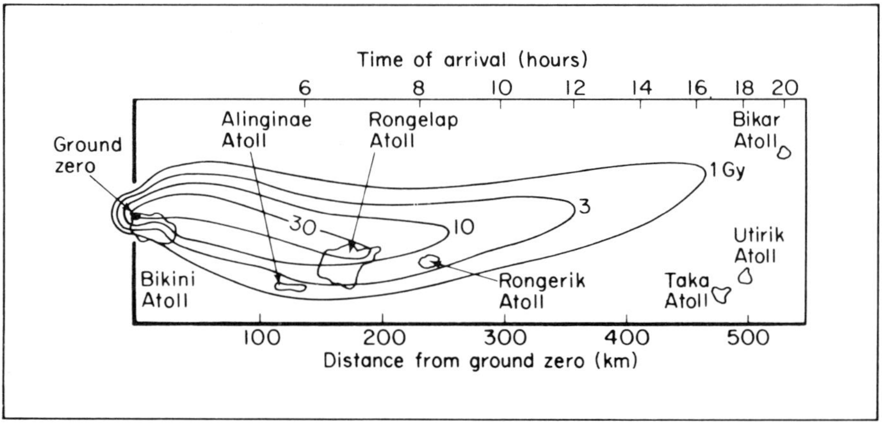

high total dose (33 Gy) at the NW tip of the Rongelap Atoll was due to a 'hot-spot'.

Two days after the test the inhabitants were evacuated. By that time some had received whole-body doses of up to 2 Gy and the majority of Rongelap islanders had beta-burns. There was also internal exposure due to inhalation and to ingestion of radioactive materials, particularly radioactive iodine, with food and water. Some inhabitants exhibited symptoms of acute exposure (anorexia, nausea and vomiting). Long-term effects, predominantly thyroid disorders, appeared later. Almost all the children had lesions of the thyroid and had to undergo surgery for the removal of thyroid nodules; later, a number of them showed symptoms of hypothyroidism. Several cases of cancer of the thyroid occurred among the female inhabitants.

The population of the Rongelap Atoll islands were allowed to return in 1957, 3 years after the test, but more than 20 years later, in 1979, the northern islands of the Atoll were still declared to be too radioactive to visit [15].

The islands of the Bikini Atoll, where testing continued until 1958, remained uninhabited for many years. Vigorous decontamination measures were taken, including the removal of 5 cm of the topsoil, before planting new trees. In 1967 the Bikini Atoll was declared habitable and some islanders returned, but later further geological surveys showed that the activity in the soil was still too high for agriculture and the Atoll was again evacuated. By the end of 1980 some islands of the Atoll were declared safe for habitation, but only if 50 per cent of the food for the inhabitants was imported [15].

4.2.3.2. Idealized pattern of fall-out

Apart from the factors mentioned earlier, the fall-out pattern is also greatly affected by irregularities of the terrain, as well as by the material of the terrain itself; for example, the particle distribution will vary according to whether the explosion occurs over a city, over fields, or over water. Coupled with the paucity of actual experimental data, all this makes the prediction of the hazard due to early fall-out highly unreliable. As an illustration, figure 23 shows a comparison of (*a*) an idealized fall-out pattern with (*b*) the pattern that might actually result from variations in local meteorological and surface conditions.

Nevertheless, an idealized picture is still useful in providing an approximate estimate of the possible radiation hazard from early fall-out. Assuming a wind of constant velocity with little directional shear, and no precipitation, the parameters of the fall-out pattern for a given reference dose-rate, namely the downwind distance, maximum width, upwind distance, and ground-zero width, can be computed for bombs of different yields exploded at the surface.

Table 16 contains results of such calculations for 1-Mt and 10-Mt thermonuclear bombs with half of the yield due to fission. The time of arrival of the fall-out, the reference dose-rates (at one hour after the explosion) and the accumulated infinity doses for persons in the open, are presented as a function of downwind direction.

Figure 23. Idealized (*a*), and possible (*b*), dose-rate contours for a 10-Mt surface burst and a wind velocity of 50 km per hour

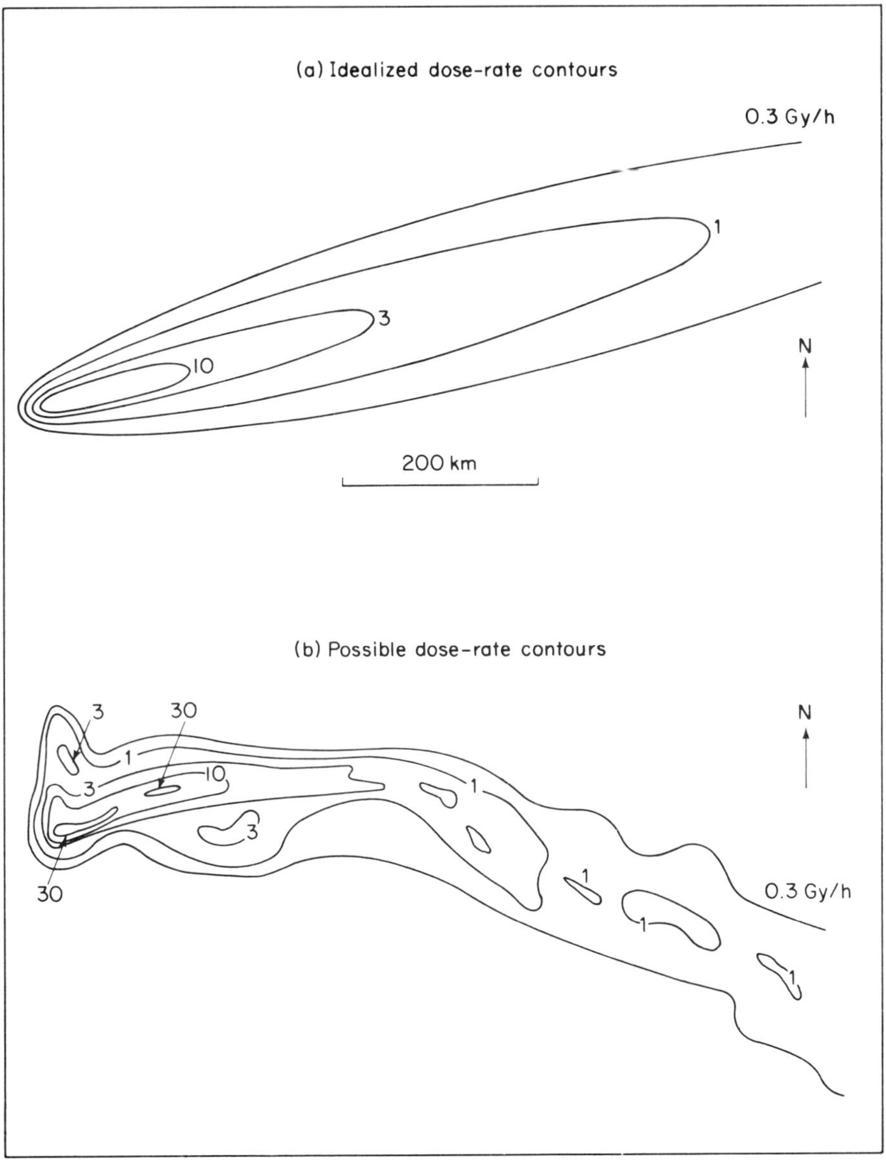

The wind velocity was assumed to be 24 km per hour. For other wind velocities the downwind distance should be multiplied by a factor. Table 17 lists the values of this factor for several wind velocities.

The time of arrival of the fall-out contains a correction for the size of the radioactive cloud, since the relevant distance is from the edge of the cloud rather than from its centre. The radii of the cloud were assumed to be 20 km for the 1-Mt bomb and 34 km for the 10-Mt bomb.

From the reference dose rates and times of arrival of the fall-out the accumulated doses to infinity (columns 4 and 7) can be calculated, using the relevant data from figure 19 (page 83).

Table 16. Dose rates and accumulated doses in an idealized pattern of fall-out

	1-Mt bomb			10-Mt bomb		
Downwind distance (km)	Time of arrival (h)	Reference dose rate (Gy/h)	Accumulated dose (Gy)	Time of arrival (h)	Reference dose rate (Gy/h)	Accumulated dose (Gy)
100	3.3	2.7	8.5	2.8	14.1	45.7
150	5.4	1.6	4.4	4.9	6.7	18.7
200	7.5	1.1	2.8	6.9	4.5	11.6
250	9.6	0.76	1.8	9.0	3.3	8.0
300	11.7	0.54	1.2	11.1	2.6	6.0
350	13.7	0.42	0.92	13.2	2.2	4.8
400	15.8	0.32	0.67	15.3	1.8	3.8
450	17.9	0.25	0.51	17.3	1.6	3.2
500	20.0	0.20	0.39	19.4	1.3	2.6
550	22.1	0.16	0.30	21.5	1.1	2.2
600	24.2	0.13	0.24	23.6	0.98	1.8
650	26.2	0.12	0.21	25.7	0.87	1.6
700	28.3	0.090	0.16	27.8	0.74	1.3
750	30.4	0.074	0.13	29.8	0.66	1.1
800	32.5	0.063	0.10	31.9	0.61	1.0

Table 17. Dependence of fall-out distance on wind velocity

Wind velocity (km/h)	Multiplying factor
12	0.75
24	1.00
36	1.13
48	1.25
60	1.38
72	1.50

4.2.4. Radiation doses from early fall-out

4.2.4.1. Maximal doses

As is seen from table 16, quite large doses, in the range of acute lethal effects, could be accumulated at considerable distances. Taking into account the variation of the LD-50 with the dose rate (see section 3.3.6) and the time distribution of the dose, one can calculate the distance at which a person in the open could accumulate an LD-50 dose. For the 1-Mt bomb this distance comes out to be about 125 km; for the 10-Mt bomb it is nearly 300 km.

If the idealized fall-out contours are considered as ellipses, with the major axis being equal to the sum of the downwind and upwind distances, and the minor axis equal to the maximum width, areas can be calculated within which given total doses could be accumulated. The area seems to be a more suitable parameter than distance, since even if meteorological conditions distort the contours, the area covered by the fall-out is likely to be less affected.

Table 18 shows the result of such area calculations for the 1-Mt and 10-Mt explosions considered in table 16. It gives the area within which the total accumulated dose could reach the value given in the first column. The areas were calculated down to a value of 0.1 Gy, which is approximately the limit of the dose which populations may accumulate in a lifetime from peace-time activities involving exposure to radiation (apart from the natural background and medical procedures). It is seen that one 10-Mt bomb could produce a dose near this limit over an area which exceeds the land area of almost every European country.

Table 18. Areas covered by given accumulated doses from fall-out

Upper limit of accumulated dose (Gy)	Area (km^2)	
	1-Mt bomb	10-Mt bomb
10	1 000	12 000
8	1 300	16 000
6	1 700	21 000
4	2 600	29 000
2	5 500	52 000
1	10 500	89 000
0.5	18 600	148 000
0.25	32 700	234 000
0.10	56 000	414 000

It will be noted that for a given value of the dose in table 18 the areas covered by the two bombs are approximately in the same ratio as the explosive yields (a factor of 10), although owing to the different heights reached by the radioactive cloud there is a systematic variation of this ratio with distance from the explosion.

From the data in table 18 the lethal area for acute radiation effects, that is, the area within which the number of survivors would equal the number of fatalities outside the area, can be calculated. For the 1-Mt bomb the lethal area is about 1 700 km^2; for the 10-Mt bomb it is about 20 000 km^2. The latter is nearly 400 times greater than the lethal area from the initial radiations (see table 12, page 70); it is also many times larger than the lethal areas for blast and heat effects.

If the population density in the fall-out area is known, one can calculate the number of persons exposed to different doses, and the possible casualties. Assuming a population density of 100 persons per km^2 (the average for Europe), about 2 million persons might receive a dose from which they would die within a few weeks or months if they were exposed to fall-out in the open from a single 10-Mt bomb. The survivors of the high doses, as well as those receiving smaller doses, would be subject to the long-term effects of radiation. From the data in table 18 it can be calculated that the collective dose (see section 3.1.1) would be about 30 million man-grays. Taking a figure of 25×10^{-3} for the probability of induction of cancer per gray (page 47), this gives about 750 000 cases of cancer, plus a somewhat smaller number of genetic defects, resulting from the explosion.

4.2.4.2. Factors affecting the dose from fall-out

The estimates made in the previous section are theoretical values and represent the maximum gamma-ray doses that a person might receive from the early fall-out. In practice, the gamma-ray doses are likely to be much smaller, due to several factors.

Because of the long delay in arrival of the fall-out in localities remote from the explosion, people would have time to leave the area threatened with fall-out or take shelter. The former alternative is of course based on the assumption that uncontaminated areas would be left and that they could be identified. This may be feasible in the case of one or a few bombs but is unlikely in the aftermath of massive bombing.

With regard to taking shelter, even though properly designed shelters would probably not be available for the majority of the population (see chapter 6), people might be expected to stay indoors, at least during the early period after the bombing. This again assumes that houses fit for people to live in would be left. Staying indoors would reduce considerably the gamma-ray dose. The dose reduction factor depends among other things on the type of the building, the floor level in a multi-story building and its proximity to other buildings. On the average, a reduction by a factor of 3 to 5 can be assumed. The number of casualties calculated in the previous section would be reduced by a factor of 5 in the case of acute effects and by a factor of 3 for long-term effects. This moderation of the hazard would of course not apply to domestic animals left in the open or to crops.

Another reduction in dose would result from weathering, that is, the removal of radioactive particles from the surface and their deposition in the depth of the soil or in water by the action of rain and wind. Weathering could reduce the concentration of radioactivity to about 60 per cent after 1 year and to about one-third in 10 years; after that the concentration would remain practically constant. Since, however, most of the dose would be delivered in the first few weeks, weathering would not contribute significantly to the dose reduction.

A greater reduction could be achieved by deliberate decontamination measures, namely, washing down or scrubbing surfaces or removal of the topsoil. However, it seems unlikely that the personnel and equipment necessary to carry out such measures would be available under nuclear war conditions.

Against these factors which may reduce the dose, one has to put the factors which may bring about an increase in the dose and in its biological effects.

The dose estimates made in the previous sections did not take into account the effects of the beta-rays, which are emitted by nearly all fission products. Beta-rays can contribute to the external hazard if the radioactive materials come into direct contact with the skin. Owing to the short range of beta-rays (a few millimetres in tissue), their action is confined to the superficial layers of the skin, but they may cause beta-burns. These start with itching and a burning sensation, and then may develop into weeping, ulcerated lesions. While causing much discomfort, these lesions are not serious and usually heal quickly; the discoloration of the skin—characteristic of beta-burns—also disappears after about a year.

The main effect of beta-rays is as an internal hazard (see section 3.7) which ensues when a person inhales air containing radioactive particles or ingests such particles with contaminated food or drink. The doses resulting from the intake of radioactive materials from fall-out depend on many factors and are not easy to calculate.

The effects due to inhalation depend markedly on the size of the fall-out particles. The nose filters out large size particles, 5 micrometres or more in diameter. It is these particles that descend first and give rise to the highest fall-out doses. For this reason, inhalation contributes relatively little to the hazard but the smaller particles which do reach the lung may settle down not only in that organ but—depending on their chemical form—also in other organs, in the bones, gastro-intestinal tract, liver, and so on.

Of greater importance is the beta-ray dose delivered via the ingestion route. While the whole-body dose resulting from the internal deposition of radioactive nuclides is a small fraction of the external dose, of the order of 10 per cent, the doses to individual organs may be as large, or even larger, than from the external gamma-rays. For example, in the Rongelap Atoll the radioactive iodine taken in the water gave rise to an internal dose to the thyroid gland nearly the same as the external gamma-ray dose, in the case of adults; in children the dose to the thyroid was 3–8 times larger than the whole-body dose.

Apart from iodine-131 and -133, the radioactive nuclides which contribute most to the internal hazard are strontium-89 and -90, yttrium-91, caesium-134 and -137, barium-140 and cerium-144. One should also mention plutonium and other actinides which are alpha-emitters and which, when taken into the body, produce much larger effective doses than beta-emitters.

An overall assessment of the effect of internal exposure indicates that for local fall-out the hazard is small compared with that from external exposure to gamma-rays as far as acute effects are concerned, but it may be comparable with the external hazard in relation to long-term effects. But even in the case of acute effects it should be noted that evidence from animal experiments (given in section 4.3.1) implies that the combination of internal and external exposures may act synergistically, namely, the LD-50 value is reduced much more than would be expected from the separate effects of the internal and external exposures.

Finally, the effect of war conditions on the chances of recovery from an acute radiation injury must be taken into account. The acute mortality rates, calculated in the previous section, were based on LD-50 values applicable to healthy individuals. With the lack of food, medical care and other social amenities, which is bound to occur in the wake of a nuclear war, many more people are likely to die following exposure to sub-lethal doses of radiation.

Taking all these factors into consideration it would appear that the numbers of casualties from the local fall-out of nuclear explosions, calculated in section 4.2.4.1, are not overestimated by more than a factor of 2.

4.2.5. Delayed fall-out

The fall-out occurring more than 24 hours after the explosion is called delayed fall-out. The division between early and delayed fall-out is an arbitrary one for surface or near-surface bursts, since the fine particles will continue to descend after 24 hours. By then the activity will have decayed to such an extent that only 20 per cent of the total dose remains to be delivered (see figure 19 page 83). Even so, considerable exposures could result from nuclides with medium and long half-lives over distances of some thousands of kilometres in the downwind direction. The greatest hazard from this fall-out arises from iodine-131, which has a half-life of about eight days; it takes about four weeks for its activity to decrease tenfold. The main route of iodine-131 into the human body is via milk from cows which have grazed on contaminated fields. The route from bomb to atmosphere, to grass, to cow, to milk, to man, is very fast. Milk with a significantly high concentration of radioactive iodine has been detected thousands of kilometres from test explosions.

A major contribution to the delayed fall-out comes from explosions which do not produce any early fall-out at all, namely, bursts at such a height above ground that the fire-ball does not touch the ground. The radioactive products are then taken up to great heights where they may become stabilized in the troposphere or in the stratosphere.

The transport of radioactive materials within the atmosphere depends on air movements, particularly at high altitudes. The knowledge of these is still incomplete, although much has been learned—as a consequence of the nuclear tests—from the study of radioactive tracers injected into the atmosphere.

The initial vertical distribution of the radioactivity within the mushroom cloud depends on the yield of the weapon. The thick and the dotted lines in figure 24 show, according to one model [16], the top and base of the mushroom cloud in equatorial and polar regions as a function of the yield.

The figure also shows the border between the troposphere and the stratosphere, the tropopause. It lies at a height of about 10 km in the polar and temperate zones and at 17 km in the equatorial zone. As is seen, weapons in the 10-Mt range, exploded at high altitude, inject all their radioactivity into the stratosphere.

The descent of the radioactive particles from the relevant regions constitutes the 'tropospheric' and 'stratospheric' fall-out; together they form the 'global' fall-out.

Figure 24. Top and base of mushroom cloud as a function of the yield of the bomb

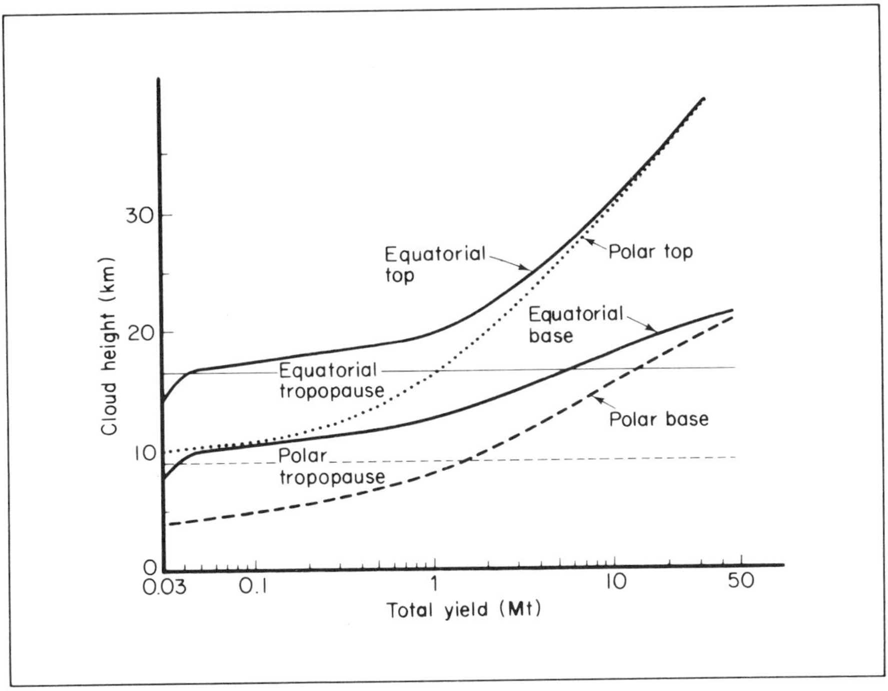

The rate of descent and the geographic distribution of the fall-out are determined by several factors, namely, the movement within the stratosphere, the transfer from the stratosphere to the troposphere, the movement within the troposphere and finally deposition on the ground. The rate at which particles are transferred from one compartment to another seems to follow an exponential law; therefore, one can speak of a 'half-residence' time, that is, the time during which half of the radioactive atoms present in a compartment are removed from it. One model [17], which predicts fairly well the atmospheric content of radioactive materials, divides the atmosphere into three compartments (see figure 25): the stratosphere above 21 km, the stratosphere below 21 km and the troposphere. The half-residence times (in months) for the transfer between these regions are indicated in the figure; for the transfer from the troposphere to ground level the half-residence time is about one month.

Within the upper stratosphere there is a fairly rapid transfer between the hemispheres. Hence, the radioactivity initially injected into the upper stratosphere will be equally divided between the northern and southern hemispheres and will eventually come down over the whole globe, although with a lower concentration in the tropical regions.

In the lower stratosphere the exchange between the hemispheres is much slower (half-residence time about five years), so nearly all the radioactivity deposited in that compartment remains in the hemisphere in which the explosion took place. The major influx of the material into the troposphere begins

Figure 25. Model for atmospheric transfer of fall-out

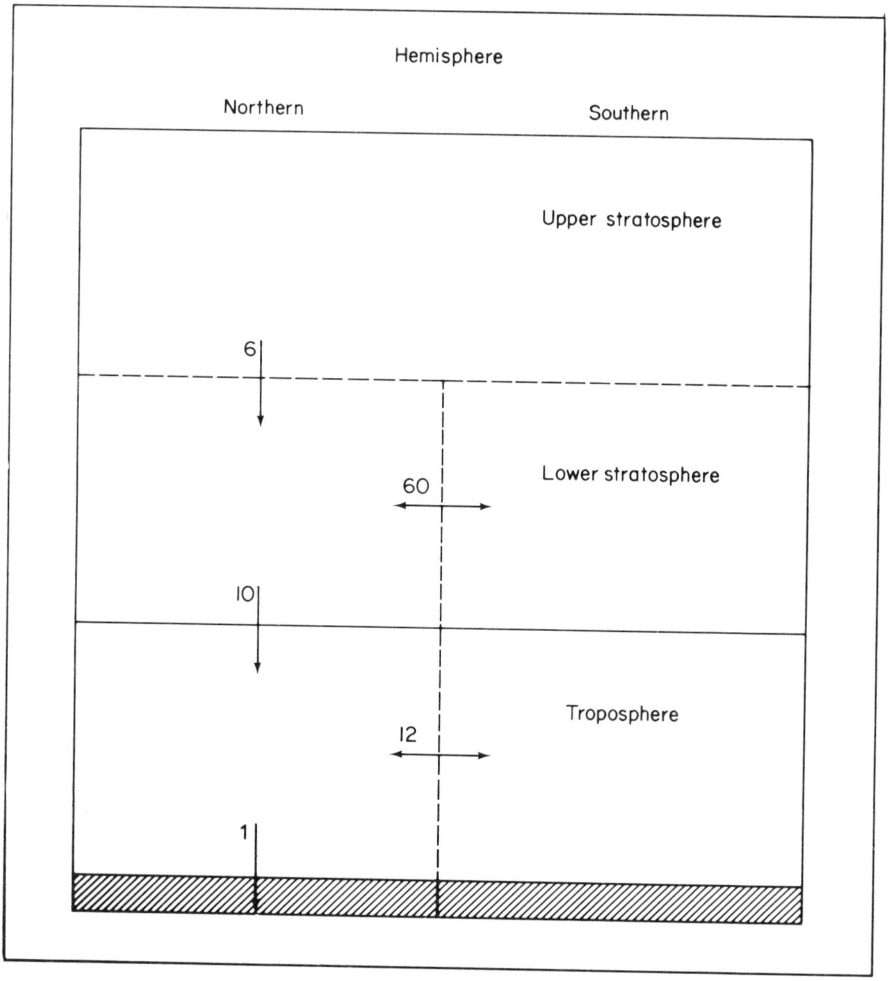

during the first winter or spring after the explosion. The activity comes down in a latitude band which spreads out by diffusion, but most of the activity remains in the temperate region.

Within the troposphere the exchange between the hemispheres is less impeded (half-residence time about one year). But, as already stated, once in the troposphere the radioactivity is deposited on the ground so fast, with a half-residence time of one month, that in practice the tropospheric fall-out also occurs in the hemisphere in which the explosion was carried out. During their sojourn in the troposphere the particles are carried round the earth, mostly in a westerly direction, and they come down in a belt of some 30 degrees in width, around the latitude at which the burst took place.

The descent from the troposphere occurs by the scavenging action of rain and snow. Dry removal by sedimentation can also occur for larger particles but it is much slower and of importance only in areas of low rainfall. The fission products which enter rain-water within the cloud give rise to rain-out (or snow-out). If they are picked up by raindrops below the cloud, they descend as 'wash-out'; the latter is a much less effective scavenging mechanism than rain-out.

The main result of the long delay in the descent of global fall-out is that all short-lived nuclides have decayed and the radioactivity, when it does come down, has become so much weaker that the external hazard due to gamma-rays no longer predominates over the internal hazard from the beta-rays. The main internal hazard then arises from the ingestion of radioactive nuclides which have entered the food chain after being deposited on the soil.

The radioactive nuclides of greatest concern in the long-delayed fall-out are strontium-90 and caesium-137. Their half-lives are so long (about 29 and 30 years) that the delay in deposition decreases their activity very little. Their fission yields are high and, moreover, since they are products of gaseous nuclides, a greater proportion of them is present in the delayed fall-out than would be expected from the fission yield. These elements are of concern because they may become incorporated in various organs of the body; strontium accumulates in bones and teeth and caesium is taken up by soft tissues. The beta- and gamma-rays emitted from them can deliver considerable internal doses. Their incorporation into the human body follows a complicated biological chain from soil and plants, to animals, to man. Strontium-90 reaches man mainly through milk and meat, while caesium-137 enters through fish, vegetables and other plants.

4.2.6. Atmospheric tests of nuclear weapons

Most of our knowledge about global fall-out is derived from the series of atmospheric tests of nuclear weapons carried out on a large scale, mainly by the United States and the Soviet Union, before the Partial Test Ban Treaty of 1963. Since then 63 further atmospheric tests (and 52 underground tests) have been carried out by China and France. The other nuclear-weapon states have carried out about 670 tests since the Treaty but they were all underground explosions. By the end of 1980 the total number of tests of all kinds was estimated to be about 1 270.

The number of tests which contributed to the global fall-out, that is, atmospheric, ground surface, water surface, and underground with venting, is believed to be about 550. The majority were fission bombs, many of a yield less than 1 kt, but the smaller number of thermonuclear bombs—ranging up to 60 Mt—contributed most of the explosive yield. The total yield was nearly 300 Mt, about half of which was due to fission.

The majority of tests took place in the northern hemisphere and therefore most of the fall-out radioactivity remained in that hemisphere (see section 4.2.5), predominantly in the temperate zone. This can be seen in figure 26, which shows the geographic distribution of strontium-90 deposited on the ground from global fall-out, as measured in 1974 [18].

Figure 26. Deposition of strontium-90 from fall-out as a function of geographical latitude

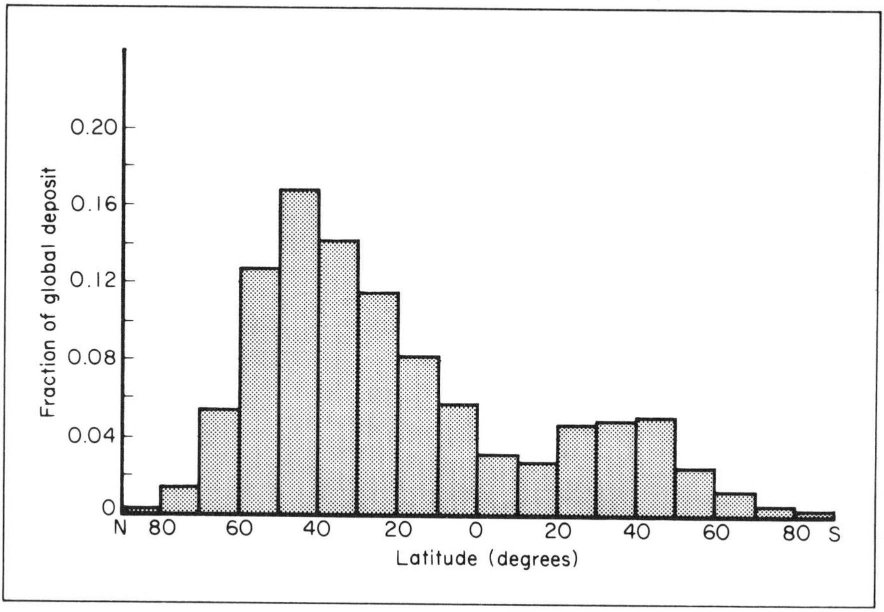

The extensive investigations carried out on fall-out from tests were aimed at calculating the dose received by man from external and internal irradiation. This involved a detailed study, for each nuclide, of the series of processes, from the injection of the nuclide into the atmosphere to its deposition in a given organ in the body. Starting from the inventory of the nuclide in the atmosphere, these processes include its deposition on the ground, uptake by plants, animals and marine biota, transfer to diet, metabolism and concentration in various organs and, finally, the calculation of the resulting dose. Not all of these processes lend themselves to direct observations. For this reason, mathematical models have to be developed and assumptions have to be made about transfer functions between different compartments and about the residence time in each of them, so as to make them fit to observed data. For example, it is extremely difficult to measure the amount of plutonium in various organs of the body; the calculation of the dose therefore depends entirely on the model used, and it may be subject to a very large error.

Table 19 is a summary of the results of these investigations for the important nuclides. They are given in terms of the dose commitment (the total dose already received plus that still to come, as the individual nuclides decay), in the same organs that were given in table 7 (page 30) for the natural background. Because of the very long half-life of carbon-14, the values given for that nuclide are the doses accumulated up to the year 2000; the later contribution from carbon-14 would add about 2 500 μGy to the dose commitment.

Because of the various assumptions made, there is considerable uncertainty about the data in table 19; for internal exposure the probable error is estimated to be a factor of 2 and for external irradiation about a factor of 5. The totals in the table were rounded to two significant figures.

The data in table 19 refer to the whole world population; the dose commitment for the population in the northern hemisphere is about three times greater than for that in the southern hemisphere.

Table 19. Dose commitment from nuclear tests up to 1975 (in micrograys)

Source of radiation	Bone marrow	Bone lining cells	Lung	Gonads
External				
Short-lived nuclides	300	300	300	300
Caesium-137	380	380	380	380
Internal				
Tritium	20	20	20	20
Carbon-14	320	290	90	70
Iron-55	4	7	7	7
Strontium-90	520	710		
Ruthenium-106			240	
Caesium-137	170	170	170	170
Cerium-144			380	
Plutonium-239		9	9	
Totals	**1 700**	**1 900**	**1 600**	**940**

Source: Sources and Effects of Ionizing Radiation; United Nations Scientific Committee on the Effects of Atomic Radiation (United Nations, New York, 1977), p. 153.

As is seen, the dose commitment from external irradiation is somewhat smaller than from internal irradiation, except for the gonads, where it is considerably larger. Strontium-90 and caesium-137 make nearly equal contributions to the dose to bone marrow and bone lining cells.

As mentioned in section 4.2.5, iodine-131 is an important contributor to the exposure of the thyroid, particularly in infants, in populations where fresh milk is a major component of the diet. It is not listed in table 19 because, having a relatively short half-life (eight days), it does not contribute significantly to the long-term dose commitment, but doses of several milligrays have been received by infants following atmospheric nuclear tests.

If we take the dose commitment data from table 19, together with the risk factors for cancer deaths (shown in table 9, page 47), it is seen that past tests of nuclear weapons are likely to give rise to nearly 100 000 additional deaths from cancer in the world. There will also be about half that number of premature deaths due to genetic damage. These are large numbers in absolute terms, but the victims of nuclear weapon testing could not be identified among the much larger number of deaths from cancer and genetic defects occurring naturally.

It should be pointed out that there has been some criticism about the validity of the models used to calculate the transfer of radioactivity from food chains to man [19]. Should this criticism prove correct, the hazard may turn out to be considerably larger.

4.3. Effects of fall-out on animal and plant life

The harmful effects of ionizing radiation described in chapter 3 apply to all living organisms, but the sensitivity to radiation varies from species to species over a wide range. Generally, the higher the species on the evolutionary scale the greater the sensitivity. In plants the sensitivity appears to be related to the volume occupied by the chromosomes in the nucleus of the cell; the larger the chromosome volume, the smaller the dose required to produce a given degree of damage. As a consequence of this, after exposure to a high dose of radiation some animal or plant species will suffer much more than others and this may seriously upset the ecological balance. For example, the killing of birds may result in a large increase in insect populations, which are much less sensitive to radiation, and this in turn would cause enormous damage to plants. A rapid increase in insect population, particularly disease vectors, may have serious effects on the health of both man and animals. Other important ecological consequences of a nuclear war may result from large-scale fires, and from the effect on the climate arising from the partial destruction of the ozone layer by the chemical reactions resulting from the explosions.

This section is primarily concerned with the consequences to man of the exposure of animals and plants to radiation, but to give an idea of the environmental damage caused even by a single bomb table 20 lists the areas of destruction of animal and plant life resulting from the explosion of nuclear bombs of 20 kt, 1 Mt and 10 Mt. For more details of the ecological impact of nuclear warfare the reader is referred to specialized publications [20].

4.3.1. Effects on livestock

The main effect of fall-out on animals is to cause external exposure by the gamma-rays emitted from radioactive substances, but the consumption of contaminated grass will in addition produce an internal exposure from the beta-rays. Beta-burns may also occur if fall-out particles remain on the skin of the animals. They may cause mucosal burns in the mouths of ruminants, leading to starvation.

As in man, whole-body exposure of animals to doses in the range of 1–10 Gy may result in death within a few weeks. The LD-50, that is, the dose which will cause 50 per cent mortality in the irradiated animals, varies between the species of domestic animals, being lowest in sheep and highest in poultry. Table 21 gives the LD-50 values for livestock; the time of death from acute effects is here taken as 60 days. The values given in the table are tissue doses received at the surface of the body; for man the corresponding dose is 4.5 Gy.

If the animals have also ingested radioactive food a smaller external dose is needed to give a 50 per cent mortality, particularly for ruminants. In experiments in which animals were exposed to both external and internal irradiation [21] it was found that an external dose of 0.7 Gy was sufficient to give 50 per cent mortality in cattle, and a dose of 2.1 Gy in sheep. Such a large reduction in the LD-50 value suggests a synergistic effect.

Table 20. Environmental damage from nuclear explosions

	Area in hectares for giving type of damage					
	Surface burst			High-altitude burst		
Explosive yield	20 kt	1 Mt	10 Mt	20 kt	1 Mt	20 Mt
Type of damage:						
Blast wave						
Craterization	1	12	57	0	0	0
Trees blown down	362	9 040	52 500	565	14 100	82 000
Vertebrates killed	24	332	1 540	43	591	2 740
Thermal radiation						
Dry vegetation ignited	749	21 300	117 000	1 170	33 300	183 000
Vertebrates killed	1 000	26 900	150 000	1 570	42 000	235 000
Nuclear radiation						
Trees killed	148	12 800	63 800	129	648	1 250
All vegetation killed	43	2 830	12 100	18	312	759
Vertebrates killed	674	36 400	177 000	318	1 080	1 840

Source: Weapons of Mass Destruction and the Environment (Taylor & Francis, London, 1977, Stockholm International Peace Research Institute), pp. 16–17.

Exposure to doses lower than lethal ones produces symptoms of radiation sickness similar to those observed in man. Damage to blood cells may result in bacterial infection and fever. Internal exposure due to the ingestion of non-soluble fall-out particles causes significant injury to the gastro-intestinal tract and gives rise to loss of appetite, diarrhoea and loss of weight. Collection of radioactive material in the rumen and abomasum causes ulceration in sheep and cattle. In cows, lactation may be reduced as a result of damage to thyroid tissue due to the ingestion of radioactive iodine. Finally, temporary or permanent sterility may be caused by radiation.

Insects are much more resistant to radiation. For example, for honey-bees the LD-50 is about 50 Gy.

Table 21. LD-50 values for livestock from external exposure to fall-out radiation

Animal	LD-50 value (Gy)
Sheep	3.5
Cattle	4.4
Swine	5.6
Horses	5.8
Poultry	7.8

Source: Survival of food crops and livestock in the event of a nuclear war, U.S.A.E.C. Symposium, Series No. 24 (NTIS, Virginia, 1971).

4.3.2. Effects on plants

The main difference between animals and plants is the larger range of sensitivities to radiation observed in plants. Different species may differ in their sensitivities by a factor of 500; if algae are included the variation may be

5 000-fold. Moreover, a given species may itself have a wide range of sensitivities, up to a factor of 50, depending on the different stages of growth. Apart from the dose itself, the effect of radiation on plants may be influenced by many environmental factors. In the case of fall-out, an additional factor is the season of its occurrence; for example, food crops irradiated in the seedling stage will be exposed for a longer time and will therefore receive a larger dose of radiation than if the fall-out occurred near harvesting time. On the other hand it would be impossible to gather in the harvest if the fall-out came down at that time.

Exposure to large doses of radiation will kill plants. The lethal doses are much higher than for animals; for food crops, even for the most sensitive plants, the LD-50 is about 10 Gy and it goes up to about 200 Gy. Estimated LD-50 values for some crops are given in table 22. At smaller doses the effects of exposure are reduced yield and height; both are dose-rate dependent, the effect being smaller the lower the rate at which the dose was delivered. Yield is more severely affected in the early reproductive stage. Plants with a growing season limited by climatic conditions may produce no yield at all, even if they survive. Flowering and ripening of fruit is delayed by exposure to radiation.

Table 22. LD-50 values for crops

Crop	LD-50 value (Gy)
Peas	< 10
Rye	10–20
Wheat, corn, cucumber	20–40
Cotton, melons	60–80
Soya beans, beet	80–120
Rice, strawberries	120–160
Squash	160–240

Source: Survival of food crops and livestock in the event of a nuclear war. U.S.A.E.C. Symposium, Series No. 24 (NTIS, Virginia, 1971).

Exposure of seeds produces mutations, most of which are deleterious.

Among trees, conifers, for example pine, are very sensitive to radiation, whereas deciduous trees are less sensitive. The LD-50 values for exposed trees range from 20 to 100 Gy. Grasses are more radiation-resistant and a dose of 200 Gy is needed to destroy grassland.

Unlike animals, where the main dose comes from exposure to gamma-rays, in plants, particularly crop plants in the early stages of growth, the dose from beta-rays may be greater than that from gamma-rays and is likely to be the dominant cause of damage. There is some evidence [21] that the combined action of beta- and gamma-rays enhances synergistically the damaging effect of radiation on crops. Even when the exposure is too small to damage the plant, the uptake by it of radioactivity from the soil may make the plant unsuitable for human consumption.

4.4. Contamination of water and food supplies by fall-out

Apart from the external exposure of gamma-rays from fission products and the inhalation of radioactive particles from the air, fall-out may give rise to internal exposure in man through the drinking of contaminated water and eating radioactive food (see section 4.2.4.2).

The main problem about water, particularly in towns, would be its availability. Supplies are likely to be interrupted by the bombing, through the destruction of equipment, pumping facilities and pipes. This means that even if reservoirs were not contaminated the water would either leak out or would not be available. Where water supplies are not interrupted, radioactive contamination is not likely to be a limiting factor. Lack of piped water, however, would compel people to seek other sources. Rain-water stored in open cisterns in fall-out areas would be highly radioactive and unsuitable for drinking, but rain falling on land undergoes a natural process of purification, so ground water would be much less contaminated. Water-softening plants—if in working order—would also reduce radioactive contamination. One of the main hazards of water in fall-out areas would be contamination with radioactive iodine, but storing the water for a few weeks would greatly reduce the dose from that source.

Fall-out may contaminate food when radioactive rain or dust settles on vegetation. The contamination could be removed by washing the food with clean water, if it was available. However, most of the contamination would come through direct assimilation of particles deposited on leaves and shoots of plants and this could not easily be removed, except by waiting for the short-lived radioactive products to decay. The same applies to milk and meat from animals which have grazed on contaminated grass. But with heavy fall-out, the long-lived products would present a long-term hazard because of the natural cycling of food chains. Radioactive vegetation is eaten by herbivores, which in turn are eaten by predators. Ultimately, both animals and plants die, returning to the soil, and are converted into humus by small animals. The radioactivity then re-enters the mineral soil. The experience of the Bikini Atoll (see section 4.2.3.1) shows how difficult it is to eliminate the fall-out particles even when vigorous decontamination measures are taken.

In the aquatic environment similar recycling may occur. Some micro-organisms have the ability to concentrate radioactivity by a factor of about 20 000 or more. These organisms are eaten by shellfish, which in turn are consumed by larger animals and eventually reach man.

The eating of animals which have been exposed to external radiation would not cause any harm from the radiation point of view, nor would it affect the food value, but if the animal had become radioactive then eating it could be harmful. Infection following radiation sickness could become a source of infection to the rest of the stock and even to man. Chickens could stop laying eggs; and when they did lay, the eggs could be contaminated. Eating stored food would be safe, but the breakdown of electricity, and consequently of refrigeration facilities, could drastically limit the availability of perishable foodstuffs.

5. Radiation casualties in a nuclear war

5.1. Calculation of casualties in a nuclear war

Under the doctrine of mutual assured destruction a nuclear war was considered to be unthinkable, because the casualties arising from it would be so great as to be unacceptable to either side. The balance of terror was thought to be the best deterrent to a nuclear war. But with the onset of new strategies, such as flexible or controlled response, the idea of a limited nuclear war began to be considered seriously. The consideration of the consequences of such a war requires making detailed calculations of the number of dead and injured under various war scenarios. Such calculations usually start out from some theoretical premise about the scale of the war and then go on to make a number of assumptions about the terrain, density of population, effectiveness of civil defence measures and so on. These data are fed into a computer program designed to estimate casualties and the result is often presented as a single total of the number of dead and injured. Such a presentation creates the impression that it is possible to calculate the casualties with some degree of precision, notwithstanding the large uncertainties in the initial conditions. Some scenarios make highly optimistic assumptions about the effectiveness of protective measures. For example, in a "limited-theatre use" of a variety of nuclear weapons, with a total yield which might exceed 1 000 kt, it was suggested that by means of such measures the total number of casualties (dead and severely injured) could be as low as 43 000 [4b]. An unintentional result of such calculations could be to make the outbreak of a nuclear war more likely, by making it appear that such a war can be waged at low cost in human lives.

A meaningful assessment of the consequences of a nuclear war cannot be made from imaginary scenarios played out by computers; instead a look at realities is required. The chief and major point to note is that in the real world the uncertainties are so large as to make any prediction valueless.

This may be illustrated by analysing the only example, so far, of the actual use of nuclear weapons in warfare. The initial conditions for the two bombs dropped on Hiroshima and Nagasaki were almost identical. In both cases fission bombs were used; compared with the present range of yields in nuclear weapons, the difference between their explosive power (12.5 kt in Hiroshima and 22 kt in Nagasaki) was negligible. They were both dropped from nearly the same height (570 and 500 metres). The targets were cities of comparable populations (256 000 and 195 000), yet the numbers of casualties were markedly different; the bomb with the smaller yield produced nearly twice as many casualties as the bomb with the higher yield, so the number of casualties per unit yield differed by a factor of 3. Moreover, the official number of total casualties (68 000 dead in Hiroshima and 38 000 in Nagasaki) are hotly disputed. Some estimates by Japanese scientists [22] put the figures at 140 000 and 74 000 respectively; these higher numbers take into account military personnel and forced foreign labour who were present in the cities. A large number of those injured by the explosions died during the next five years, but there is great uncertainty about the actual number since no census of the population was taken until 1950. We shall probably never know how many casualties there actually were.

It should also be noted that fall-out—the hazard of which it is even more difficult to estimate—was almost completely absent in the two explosions.

Thus, even in the only instance in which nuclear weapons were employed against a population, an event which was subsequently investigated very thoroughly, large uncertainties remain about the absolute numbers of casualties; the event also demonstrated that there can be a wide spread in the number of immediate casualties per unit of explosive yield. How much greater must be the uncertainty in a hypothetical war!

5.2. Uncertainties in estimates of total casualties

The factors which contribute to the uncertainty can be considered under a number of headings.

1. Postulated initial conditions. Leaving aside the case of an all-out war, in which all the stockpiles of nuclear weapons might be used, the scenarios usually start with a postulated scale of the war, with the implied assumption that there will be no escalation. The postulates include the type and number of weapons, the CEP, and the mode of explosion, that is, the height of bursts. A range of values of total explosive power is usually postulated, which brings in an uncertainty even in the initial conditions, quite apart from the uncertainties attached to the accuracy and reliability of the weapons.

2. Explosion parameters. The factors contributing to the uncertainties include the explosive yield of each weapon, the fraction of fission in it, the materials used in the assembly of the bomb and the height of burst. The bomb may not necessarily achieve the designed explosive power. Although it is conventional to assume that in F-F-F bombs half of the yield is due to fission, the actual proportion can vary over a wide range, with a corresponding uncertainty about the fall-out intensity. Depending on the materials used in the assembly, the intensity of the initial radiations (neutrons and gamma-rays) may vary by more than an order of magnitude. A weapon intended to explode at a certain height in the air may in fact detonate nearer the ground or even on the surface, thus producing intense local fall-out. In any case, surface detonations, by contact fuses, may be preferable, even though they reduce the area of blast damage, to avoid possible jamming of radio-controlled fuses.

3. *Target parameters.* It is usual to specify whether the targets are centres of population or military/industrial installations. However, the latter are often situated on the outskirts of cities, and an error in targeting may result in the centre of the city receiving the hit, with a vast increase in casualties. The reverse situation might also occur with a bomb aimed at the centre of a city landing on the outskirts. The type and structure of buildings may greatly affect the magnitude of blast effects as well as the probability of starting fire storms or conflagrations. The nature of the terrain may provide different shielding factors against blast or radiation effects.

4. *Seasonal parameters.* The time of day: an attack in daytime, when people are at work in the centre of cities, will produce many more casualties than an attack at night, when people are at home, since many live in the suburbs. The time of year: an attack in winter may cause more casualties from cold and exposure among those forced to leave their homes, and greater difficulties in providing fall-out protection; an attack in spring may produce lethal irradiation of plants resulting in famine; an attack in summer may prevent the gathering of the harvest.

5. *Atmospheric parameters.* Weather: dry weather may spread fires and lead to a conflagration; rainy weather may cause precipitation of fall-out. Wind conditions, which also vary with the season, make all the difference to the distance and location of fall-out deposition. Good visibility greatly extends the area of lethal thermal radiation.

6. *State of preparedness.* There is clearly a large difference between a war starting with a surprise attack, or being preceded by a period of a gradually deteriorating political climate. Would governments give early warning, with the inevitable economic and societal upheavals, which might turn out to have been completely unnecessary? Or would governments try to keep the population calm by presenting the probability of a war as being small, thus risking being caught unready with regard to measures requiring early action?

7. *Post-war societal conditions.* The number of casualties *after* the actual war is over is influenced by many factors, such as the general state of health and hygiene of the population, shortfall of medical care, the likelihood of epidemics, lack of food, shortage of housing, failure of communication, economic and social disruption, and breakdown of governmental control.

8. *Synergistic effects.* Finally, the synergistic effects arising from all the aforementioned factors cannot be overstated.

Some of the above parameters are under the control of the attacking side. But wars never go exactly according to plan, and mistakes due to human failure or technological malfunction are inevitable. Other parameters are outside the control of either side and their effects are quite unpredictable.

This is by no means an exhaustive list of uncertainties, but it suffices to prove that any projected estimate of the total casualties in a nuclear war which pretends to be accurate to better than an order of magnitude must be viewed with much scepticism.

5.3. Uncertainties in estimates of fall-out casualties

Of the various types of damage resulting from a nuclear war, those relating to radiation fall-out are the most difficult to predict and to estimate, since so much is outside the control of the attacker and the attacked. The unpredictability of the weather is a major factor in the uncertainties. If a surface burst is intended to destroy a military target, then depending on whether the wind happens to blow in such a direction that the local fall-out will embrace a large city or open country, there could be an enormous difference in both the number of people receiving lethal doses of radiation and the number receiving smaller doses with resulting long-term effects. The numbers will also greatly depend on other meteorological conditions, the type of cloud formation, the temperature, and whether there happens to be rain or snow at the time. Rain falling later may wash away some of the fall-out material and deposit it elsewhere. The fall-out distribution depends also on the nature of the terrain, and, as was shown in section 4.2.3.2, its pattern is highly unpredictable.

The experience of the Marshall Islanders, following the Bravo test of 1954, illustrates the unpredictability of the fall-out hazard. The fact that after 27 years the soil on some parts of the Bikini Atoll is still considered to be too contaminated for permanent habitation is a reminder that one cannot put too much reliance on the rapid decay of the fall-out radioactivity. Standards of radiation safety are bound to be lowered in wartime, and people will be coming out from houses or shelters and spending time in the open when the dose levels will still be so high that they would be considered unacceptable in normal conditions; but the inevitable long-term effects will still take their toll.

Because of the persistence of radioactivity after the attack is over, the hazard from fall-out is most sensitive to the degree of preparedness to deal with it. The difficulties associated with civil defence measures will be discussed in chapter 6. In view of the lack of civil defence measures in many countries and the lamentable ineffectiveness of measures taken in other countries, the assumption in some analyses that civil defence measures will reduce the casualties by a considerable factor is clearly unjustified.

Finally, there is the uncertainty in the absolute magnitude of the long-term effects of radiation, for example the number of deaths from cancers and genetic defects which may result from exposure to a given dose of radiation. As explained in chapter 3, this alone may introduce an uncertainty by an order of magnitude.

5.4. Methods of presenting the consequences of a nuclear war

When all the uncertainties listed in sections 5.2 and 5.3 are taken into account, it is clear that a single figure for the number of casualties in a given war, calculated as an average value, is quite meaningless and could be grossly misleading. How then are the consequences of a nuclear war to be presented?

One method would be to use a statistical approach. The argument here is that the likelihood is very small that *all* the factors contributing to the casualties will act in the same direction, and that it is more likely that only some of the parameters which result in an increased number of casualties would be operative. By calculating the probability for a given number of events to occur simultaneously, one can obtain a frequency distribution curve as a function of the number of casualties. Such a curve would have a peak, which would give the median number of casualties; the spread of the curve would allow an estimate to be made of the range of values for a given level of probability of occurrence. From what was said above it is clear that such a curve would be highly skewed on the side of high casualties, and it might follow a log-normal distribution, that is, one in which the number of casualties is more suitably plotted on a logarithmic scale. The range of values would then be given as a lower and upper multiplier of the median value. For example, an assumed war might be calculated to result in a median value of 10 million casualties, with a 90 per cent probability that the actual number will lie within a factor of 5 of the median, that is, between 2 million and 50 million.

An example of a probability computation of the number of casualties presented in a somewhat different way from the above is shown in figure 27. It gives the results of a Rand Corporation computer program [23] to calculate the fatalities in a nuclear war postulated to be limited to a Soviet attack on all US ICBM silos. In figure 27a, the horizontal scale gives the number of fatal casualties and the vertical scale the calculated probability that the given number of deaths would lie within certain limits. The skewed distribution is clearly seen, since the number of fatalities is plotted on a linear scale. The median value is 820 000 deaths, and there is a 90 per cent probability that the actual number would lie within a factor of 2.2 of this median, that is, between 400 000 and 1.8 million.

Another way of presenting the same results is shown on figure 27b. The vertical scale now gives the cumulative probability, that is, the probability that the total number of casualties may reach the given value on the horizontal scale. Thus, it is seen that there is a 50 per cent probability that the number of deaths may be up to 800 000. There is a nearly 100 per cent probability that the number of fatalities will be less than 3.3 million.

The data in figure 27a and b apply to specific values of the various parameters (explosive yield, height of burst, fraction of fission, characteristics of the wind). Changes to some of these parameters, fed into the same computer program, led to considerable changes in the probability curve. Figure 27c shows three such variants. Curve 1 illustrates the same situation as in figure 27b; curves 2 and 3 were obtained for different values of the parameters. It is seen that even for the same type of postulated attack, a Soviet counterforce attack on the ICBMs in the United States, a very wide range of fatalities can be calculated, from 330 000 to 18.2 million.

A different approach to estimating the number of casualties is that of the 'worst case analysis'. In this method the maximum number of casualties is calculated on the assumption that all factors which can contribute to increasing damage will operate and should be taken into account. For example, in figure 27, the number of casualties according to the worst case analysis would be given by the value for nearly 100 per cent probability, that is, 18.2 million. It should be noted, however, that an even larger number of deaths could be calculated if more pessimistic assumptions were made about the parameters.

Which of the two methods of estimating the casualty toll is adopted depends on basic concepts about nuclear war, namely whether a nuclear war is regarded as unthinkable, because once started it is bound to escalate into an all-out war, or whether it is accepted as a fact of life in modern society.

Figure 27. Computation of fatalities from a counterforce attack

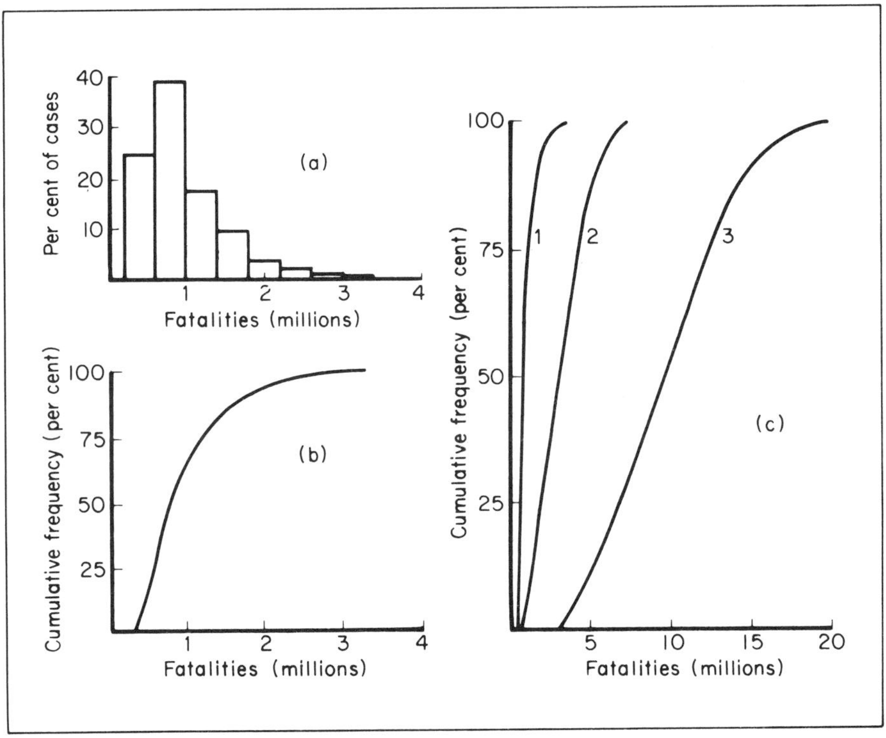

With the first concept the statistical approach is obviously unsuitable, because should a nuclear war occur it would be a one-off event, perhaps even an end-all event. Those adhering to the second concept would say that the evaluation of the *probable* results of a nuclear war is as necessary as the evaluation of any other untoward occurrence resulting from human activities.

This brings us back to the problem of a limited nuclear war and the likelihood of containment of the use of nuclear weapons. Should nuclear weapons become available to many nations, it is conceivable that a country with a very small nuclear arsenal might be driven by desperation to attack another country with a similar small stockpile, or without any nuclear weapons. If both these countries were non-aligned with the big-power blocs, a one-sided attack or an exchange of nuclear weapons between them might possibly remain an isolated event. But in any situation involving aligned countries, or when the security or interests of the major powers are affected, such as in Europe or the Middle East, the probability of isolating the conflict is greatly diminished. In particular, a clash on the European theatre, initially intended to be limited to the use of tactical nuclear weapons on military targets, is likely to escalate, first by nuclear attacks on cities in the theatre zone, still with purely military aims in mind, and then by attacks on targets in the homelands of the adversaries. Eventually the war would escalate to a full-scale nuclear exchange, with the employment of all the stockpiles of nuclear weapons. It is impossible to evaluate numerically the probability of such an escalation, but it is certainly significant. In any case, the Soviet Union has stated that its response to *any* use of nuclear weapons would be massive retaliation. If one therefore accepts the view that escalation is probable, only the consequences of an all-out nuclear war need to be considered and the worst case analysis appears to be the appropriate approach.

5.5. Casualties in an all-out nuclear war

In an all-out nuclear war the assumption is made that all nuclear weapons in the stockpiles will eventually be used. These stockpiles have been increasing over the years, although with the improved accuracy of missiles the yield of the weapons has tended to decrease. This has been so for the arsenals of the United States but the Soviet Union is following suit. One result of using weapons of lower yield is much less 'overkill' and a greater destructive effectiveness per unit yield. As discussed in section 2.3.2, the total explosive yield in the nuclear weapon inventories at the present time may amount to 20 000 Mt, and this megatonnage will be used as a basis for estimating the casualties in an all-out nuclear war.

A recent publication of the US Office of Technology Assessment (OTA) contains calculations of casualties from nuclear attacks according to several scenarios [3b]. One of these considers a massive Soviet attack on US military and economic targets plus a somewhat smaller retaliatory attack on Soviet targets; these conditions come near to an all-out nuclear war. A total of 7 800 Mt was assumed to have been used by both sides: 6 500 Mt on the USA and 1 300 Mt against the USSR. The number of immediate deaths from blast, heat, initial radiation and early fall-out was calculated to reach 165 million in the United States and up to 100 million in the Soviet Union. In calculating these numbers it was assumed that all weapons were exploded at the surface and that no civil defence measures were taken; however, the shielding provided by walls of buildings for those staying indoors was taken into account.

Although the attacks were assumed not to be aimed specifically at cities or other centres of population, nevertheless most cities were shown to be destroyed and the majority of their inhabitants killed. Local fall-out played a big part. As an example of the latter, the shaded part of the map of the United States on figure 28 shows the area in which persons in the open would receive a dose of 1 Gy or more; it covers about 80 per cent of the country [24].

Figure 28. Area of the United States with potential radiation level greater than 1 Gy

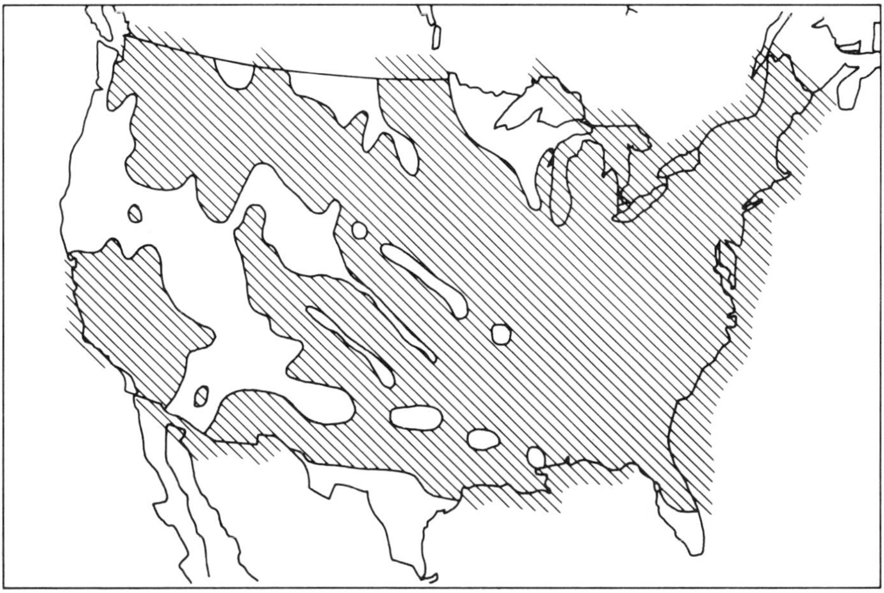

In addition to acute deaths and injuries there would also be long-term effects, both inside and outside the combatant countries. Table 23 summarizes the upper limits of the different types of casualty, according to the OTA scenario. The overall total number exceeds 500 million; the largest item is, of

Table 23. Possible total casualties in an exchange of 7 800 Mt between the USA and the USSR (OTA scenario)

	In the combatant countries (millions)	In other countries (millions)
Immediate casualties		
Number of deaths	265	
Number of injured	133	
Long-term casualties		
Cancer deaths	14.8	9
Thyroid cancers	7.5	3.2
Thyroid nodules	10.9	9.5
Abortions	14	5
Other genetic effects	21.5	15

course, the number of immediate deaths. An OTA variant of the above scenario, to include attacks on populations, shows that the number of immediate fatalities could go up to 320 million, that is, about 70 per cent of the combined populations of both countries (88 per cent in the USA alone).

All these numbers refer to casualties following the physical destruction wrought as a result of the explosions. They do not include the casualties which might result subsequently from social upheavals in the aftermath of war, from epidemics, lack of food and armed combat between rival factions struggling for survival or hegemony. Eventually an ordered society may be restored, but the damage to the economy is likely to be permanent. In the words of the OTA Report "... the sheer magnitude of such an attack would place in question whether the United States would ever recover its position as an organized, industrial and powerful country". The same would presumably apply to the Soviet Union.

The OTA scenario did not take into account some other possibilities which might lead to an increase of the effect of local fall-out and impede recovery after the war.

One is the consequence of attacking nuclear energy facilities, namely power reactors, storage tanks and reprocessing plants. In an attack aimed at damaging the economic strength of a country, nuclear power stations are bound to be prime targets. If these facilities were attacked with nuclear weapons, even of low yield, their immense radioactive contents would be released into the atmosphere, carried with the mushroom cloud, and deposited with the local fall-out. However, there would be a big difference from ordinary fall-out: the rate of decay of the radioactivity from reactors and other nuclear facilities is much slower (see section 7.1). This means that large areas of land would remain radioactive and uninhabitable long after the fall-out from the bombs had decreased to tolerable levels. Even with the present, quite small number of nuclear reactors, almost all of Central Europe would be covered by the slow-decaying fall-out from these reactors.

Another, somewhat more speculative possibility is that if a number of nuclear weapons are detonated at about the same time, the radioactive plumes from the individual clouds may interact with each other to cause a larger proportion of the fall-out to descend than is usually assumed for local fall-out from single bombs.

The OTA scenario will now be extended by assuming that all the 20 000 Mt of the present inventories will be exploded, and that the war arena will include the whole of Europe. Scaling up the casualty toll listed in table 23, but taking into account the much higher density of the population in Europe, leads to an estimate of over 80 per cent of the population in the countries involved killed and 15 per cent injured. This means that some 800 million persons may die from immediate effects; many of the 150 million injured may also die within a few months.

The casualty toll might be even higher if the war arena were to include some Asian countries, such as China and Japan.

Thus, the immediate outcome of an all-out nuclear war might be the loss of one-fifth to one-third of the world population. Moreover, the devastation of nearly all the industrialized countries which provide much food, fertilizers and other vital aids to the underdeveloped countries, in many of which the population is on the brink of starvation, could drastically affect the Third World. How many deaths would result from the ensuing famine and other deprivations in those regions is impossible to say, but the total may run into hundreds of millions.

Apart from these indirect effects, the non-combatant countries will also be affected by the long-term effects of global fall-out. A simple scaling up of the casualties in the non-combatant countries given in table 23 leads to about 100 million further casualties from somatic and genetic effects of radiation. Other effects, not included in the OTA scenario, are skin cancers and melanomas from the increased ultra-violet radiation following partial depletion of the ozone in the atmosphere.

As was emphasized many times, all these estimates—even though based on the worst case analysis—carry many uncertainties, and could be wrong by a large margin in either direction. But, whatever the detailed calculations of the direct and indirect number of casualties in an all-out nuclear war, several predictions can be made with confidence: there would be no victors in such a war; no political system or ideology would survive it; and no ideals or ways of life for which the war may have started would prevail. Those in the combatant countries who managed to escape alive, as well as the populations in other countries, would find themselves in a new and strange world in which physical survival would be the main driving force for a long time.

5.6. *Maximal effect of radiation*

The previous section has shown that even in an all-out nuclear war the majority of the world population would survive physically, although civilization as we know it would be destroyed. On the other hand, statements have often been made that a nuclear war would wipe out not only civilization but also human life on this planet. With the present stockpiles of nuclear weapons and the likely method of their employment, such statements are clearly not true. But the margin is not very large. Should nuclear weapons become widespread among nations and should the total arsenal go up by a factor of 10, an extension of the calculations made in the previous section would lead to the conclusion that the total casualty list might embrace all human beings. Most of the toll would be from the immediate effects of the blast, heat and initial radiation, but the exposure to the radiations from fall-out could by itself lead to the same result.

An analysis of the types of weapon in the present stockpiles shows that of the 20 000 Mt, a yield of about 13 000 Mt would be due to fission. Let us start with a purely imaginary assumption that the fission products from all this megatonnage are uniformly distributed over the land area of the globe (1.5×10^8 km^2) near the surface, so as to create the maximal conditions for fall-out. As already stated (section 4.2.1.6), the radioactivity from a fission bomb of 1 kt could produce a gamma-ray dose rate, at one hour after the explosion, of 75 Gy per hour, on a surface area of 1 km^2. Therefore, the 13 000 Mt could give rise to a dose rate of 6.5 Gy per hour, at one hour, over the land area of the globe. The total accumulated dose by persons in the open, due to fall-out alone, could thus be about 60 Gy, which far exceeds the value for early death from the acute effects of radiation.

However, a number of corrections must be applied to this figure which will reduce it considerably. Only about 60 per cent of the fission-product content of the bomb appears in local fall-out and only 70 per cent of that is effective, due to unevenness of the ground. Above all, the radioactive particles in the fall-out are not distributed uniformly; their concentration falls off rapidly with distance and the dose rate decreases due to the decay during the time it takes the fall-out to reach the ground. For these reasons, the calculation of the area covered by a given dose has to be made by the method used in section 4.2.4.1. With this method, the actual size of each detonation does not affect the result significantly, but the scenario requires that the explosions take place in a certain pattern.

With these assumptions the calculations show that the smallest infinity dose that a person might receive from gamma-rays, if he stayed in the open, is about 4 Gy. Within a factor of 2, this is a dose that might cause 100 per cent mortality when taking into account the effect of dose rate. Considering the large uncertainties in all such calculations, and adding the effects of internal exposure and the other war-condition parameters which tend to lower resistance to radiation, it seems that—at least in theory—radiation alone might threaten the survival of mankind, even with the present nuclear arsenals.

Should the arms race continue with an increase in stockpiles by a factor of 10—particularly if more nations accumulate nuclear weapons—the above scenario will tend to become more realistic; even with the explosions occurring randomly, and allowing for some protection by staying indoors, the likelihood of everybody accumulating a lethal dose would become quite high. A further increase of the stockpiles by a factor of 10 would make it a certainty, with the global fall-out assuring the death of those who survived the local fall-out.

An increase in the stockpiles by a factor of 100 seems fantastic, but who would have believed in the mid 1950s, when the nuclear arms race began, that the world would need and accept a more than 100 times greater nuclear stockpile by 1980? Unless there is a halt to the arms race it is likely to go on by its own momentum and increase both vertically and horizontally. We may then reach the stage when the "On the Beach" scenario [25] will no longer be a subject of fiction.

6. Effectiveness of civil defence

Defence measures against a nuclear attack can be either active or passive. Active measures aim at preventing the nuclear weapons from reaching their targets, for example by destroying the offensive missiles in flight. The passive approach aims at minimizing the effect of an attack by two kinds of measure: military, to make some targets less vulnerable, and civil, to reduce the number of casualties and the economic damage.

The effectiveness of all these types of defence has been extensively examined and debated. The next section reviews the military measures, and it will be shown that these are largely ineffective and generally lead to a stepping up of the arms race. The rest of the chapter discusses civil defence measures; although this subject is much more controversial, it seems that similar conclusions should be drawn about the effectiveness of civil defence as about military defence measures.

6.1. Military defence

Among the active measures of defence, the main emphasis in the early years was on anti-ballistic missile systems (ABMs). The aim of such a system is to detect, identify, intercept and destroy in flight the warheads of the attacker's missiles. The destruction could be planned to occur either before or after the re-entry of the offensive missile into the lower atmosphere. For this purpose, the defending side would fire a nuclear missile into the projected path of the hostile missile and explode it within such a distance that the effect of the explosion would either cause the physical destruction of the hostile missile by the blast wave or—more likely—make the guidance or detonating mechanisms inoperative by the effect of the X-rays produced at the instant of the explosion, or by the action of the neutrons, the gamma-rays or the thermal radiation.

ABM systems began to be deployed in the USA and the USSR in the early 1960s, but soon their limitations and detrimental consequences became obvious. It is relatively simple to make ABMs ineffective by the use of decoys, that is, objects looking like missiles, a number of which could be fired by the attacker at the same time as the real missile, thus saturating the defence system. Indeed, MIRVs were initially developed as decoys for ABMs. However, the main concern was that the response to the introduction of ABMs would be an increase in the number of attacking missiles, so as to compensate for any losses. In a race between the offensive and defensive strengths the advantage is on the side of the offensive, particularly since in most cases the cost of an increased offensive strength is much smaller that the corresponding cost on the defensive side. These arguments were well recognized and have led to the SALT I agreement, in 1972, limiting the number of ABM systems to be employed by the USA and USSR. In fact, even the options allowed by the agreement were not taken up.

With the end of ABMs attention began to be focused on other ways of destroying hostile missiles, for example by means of directed-energy weapons, that is, collimated beams of high-energy charged or neutral particles produced in accelerators or powerful laser beams. Some military sources in the USA and the UK have alleged that the USSR has developed particle accelerators for such purposes. But on purely physical grounds it can be shown that charged particle beams from accelerators based on the ground would be completely ineffective; the same applies to neutral particle beams. The development of high-energy laser technology may make the use of lasers theoretically possible, but several other considerations, including the cost of such systems, make them impracticable. Nevertheless, attempts to develop this technology are likely to be intensified in the near future, with the only certain result that they will contribute to a further destabilization of the situation.

Among the passive methods of defence, one is hardening of silos, that is, putting the ICBM at a greater depth in the silo and protecting it with a greater thickness of concrete. However, with the much increased accuracy of present day missiles the blast effect from one or two nuclear warheads would cause enough structural damage to the silo to prevent the firing of the missile in it. A more effective way would be to make the ICBMs mobile, as in the projected MX system (described in section 2.3.2). But as was pointed out earlier, the likely response to such measures is an increase in the offensive strength.

To sum up, military defence measures are either ineffective, or very expensive, or both. In the absence of arms limitation agreements they are likely to lead to a further increase in the nuclear arsenals.

6.2. *Civil defence*

Civil defence measures aimed at reducing the number of casualties in a nuclear war can take different forms. As far as radiation is concerned the two main approaches are physical protection, in special shelters or in basements of buildings, and evacuation from fall-out areas. In both cases, the effectiveness will depend on the scale of the war and the degree of preparedness well in advance of it. There can be no doubt that with an adequate civil defence organization numerous lives could be saved in a limited nuclear war. But the larger the scale of the war the less is the effectiveness of any civil defence measure and—as was the case with military defence—if the main aim of the enemy is the destruction of the population, for example in revenge for an attack on his population, then the severity of the attack can be scaled up to match the civil defence measures and thus neutralize them to a large extent. In view of this, the decision to set up a civil defence system is by no means clear-cut, which partly explains why in most countries the civil defence issue has so far not been seriously tackled.

The three main factors which militate against the institution of large-scale civil defence measures are: (*a*) capital expense—an adequate civil defence system would be extremely costly and would add significantly to the economic burden of the country; (*b*) the huge effort needed to set up a special organization and to train and keep in employment a very large staff; and (*c*) preparation of the population to face the possibility of a nuclear war.

This brings us to the crux of the matter: the posture in relation to a nuclear war. All nuclear weapon states maintain that they need nuclear weapons in order that they will not be used, that is, to deter an adversary from using theirs. If this doctrine of mutual deterrence were taken seriously, civil defence measures would not be necessary; indeed, if both sides adhered to the doctrine of mutual assured destruction, the deterrent would be more effective if no civil defence measures were taken. But the argument loses its strength with the counterforce, or limited war, concepts, under which the outbreak of a nuclear war is considered feasible. With this notion comes the need to take steps to reduce collateral damage, although the very fact of taking measures to reduce the casualty toll, thus making a nuclear war more acceptable, might by itself make the outbreak of such a war more likely.

In countries where governments have to publicly justify budgets and policies, they are placed in an embarrassing situation in relation to civil defence measures; this applies in particular to the nuclear weapon states where vast sums are spent on maintaining and up-dating nuclear armaments. Justification for this expenditure is sought on the grounds that it will ensure, by virtue of deterrence, that a nuclear war will not happen. But if the probability of a nuclear war is presented as being low it becomes difficult to justify an enormous expense on civil defence, that is, on preparation for an event that is unlikely to happen.

On the other hand, if the official policy of using nuclear weapons in certain events (for example, the NATO policy to respond with nuclear weapons to a Soviet attack in Europe with conventional weapons) is to be credible, governments must show that they can cope with a limited nuclear attack by having taken steps to protect the civilian population.

The result is a compromise, by which the governments are more likely to succeed in bluffing their own citizens than the enemy. The citizens are encouraged to take some simple measures and are told that these recommended measures will drastically reduce the casualty toll in a nuclear war. Some of the official literature, apart from underestimating the consequences of such a war, contains factually incorrect information. For example, the pamphlet "Protect and Survive" issued by the British Government [26] starts with the following statement: "The heat and blast are so severe that they can kill, and destroy buildings, for up to five miles from the explosion". By saying "up to five miles" (8 km) it implies that there would be no deaths beyond that distance. A comparison with the data in figure 3 (p. 15) shows this to be false in relation to most weapons in the Soviet strategic nuclear arsenals.

The only countries that have taken civil defence measures seriously, and have instituted a systematic shelter programme, even if not specifically designed for nuclear warfare, are the neutral states Sweden and Switzerland. They do not expect a direct nuclear attack, but are concerned about the fall-out which may reach their territories from nuclear attacks on neighbouring countries.

The problems which may be encountered in attempts to provide an adequate civil defence system are discussed below, after a description of the physical aspects of protection against radiation. The conclusion that emerges from a consideration of these problems is that civil defence measures could significantly reduce the number of casualties in some types of war, in some countries, under some conditions. But they would become progressively less effective as the scale of war increased and would be quite ineffective with an all-out nuclear war. In view of the high probability of any nuclear war escalating into an all-out war, it is very doubtful whether the expense and effort in preparing a civil defence system for a nuclear war are at all worthwhile. This is particularly so if it is accepted that the initiation of civil defence measures may itself contribute to the likelihood of a nuclear war.

6.3. Attenuation of radiation

With external sources of radiation, any material interposed between the source and the human body will reduce the dose to tissues. The degree of attenuation depends on the type of the radiation and the absorbing material. Alpha-particles—say from plutonium or other actinides—have a very short range, and unless the alpha-particle source is very close, within a few centimetres of the skin, they will not produce any external dose. Beta-particles, emitted from most fission products, have a longer range, about 1 metre in air, but a thickness of a few millimetres of most materials will stop them. Ordinary clothing is sufficient to protect the body from alpha- or beta-particles, but care must be taken that no radioactive material on clothing comes into contact with the skin, nor is taken into the body by inhalation or ingestion. A shower, plus a change of clothing, or a thorough rinsing of the clothing are necessary and sufficient measures. However, this might not be possible under war conditions, and the resulting internal hazard (see section 3.7) could prove to be serious.

Protection is much more difficult with neutrons and gamma-rays, since these types of radiation do not have a finite range. Theoretically their intensity never reaches zero, but in practice it can be reduced to a negligible value. The degree of attenuation depends on the thickness of the material between the source and the body; the intensity of the radiation which penetrates the material decreases with thickness approximately in an exponential manner. For a given material one can therefore state the thickness which will reduce a given radiation by a certain proportion, say by a factor of 10. For example, if 11 cm of iron reduce the intensity of the initial gamma-rays to one-tenth, 22 cm of iron will reduce it to one-hundredth and 33 cm to one-thousandth.

Table 24 gives the tenth-value thicknesses for some relevant materials for three types of radiation. The difference in attenuation between gamma-rays from the initial radiation (source 4 in table 10, p.63) and those from fall-out is due to the different initial energies. For the neutrons an initial energy of 0.3 pJ was assumed.

Table 24. Tenth-value thicknesses for radiation (in metres)

Material	Density (kg/m^3)	Neutrons	Initial gamma-rays	Fall-out gamma-rays
Iron	7 860	0.08	0.11	0.084
Concrete	2 350	0.24	0.41	0.28
Earth	1 600	0.32	0.61	0.41
Water	1 000	0.14	0.99	0.61
Wood	640	0.18	1.60	0.97

In the case of gamma-rays the attenuation by a given thickness varies approximately in inverse proportion to the density of the material, so that the product of the tenth-value and the density is nearly constant for all materials. It can be seen from table 24 that this product is about 970 and 640 kg m^{-2} respectively for the two types of gamma-rays. From these, the tenth-value for other materials can be calculated. For example, since the density of lead is 11 300 kg m^{-3}, the tenth-value of lead for the initial gamma-rays is about 0.086 metres.

This simple relation does not hold for neutrons, because the processes of attenuation of neutrons are different from those of gamma-rays. Heavy- or medium-weight elements attenuate the neutron energy by inelastic collisions, whereas light elements, in particular hydrogen, are more effective in slowing down neutrons to thermal energies by elastic collisions. Thermal neutrons are very easily absorbed in many materials, for example boron or cadmium, but there is no correlation with density.

The inelastic collisions of neutrons give rise to gamma-rays, so even if one starts with a pure neutron source, it is increasingly accompanied by gamma-rays as the neutrons pass through matter. The most effective attenuation of radiation, be it neutrons or a mixture with gamma-rays, is achieved by a combination of light- and medium-weight elements. Concrete, which is such a combination, is therefore the best material.

6.4. Protection from radiation

From data such as those given in table 24 one can calculate the protection factors offered by walls of buildings or different types of shelter. The protection factor is the ratio of the intensity of the radiation before and after the passage through the protection structures. Table 25 lists protective factors for a variety of structures. For the reason stated above, the best protection, for a mixture of neutrons and gamma-rays, is provided by concrete, particularly heavy concrete which contains iron or barium.

Table 25. Radiation protection factors

Structure	Neutrons	Initial gamma-rays	Fall-out gamma-rays
Frame house	1.2–3	1.2–1	15–30
Basement	1.2–10	1.7–10	10–20
Multi-storey building			
Upper storeys	1–1.1	1.1–1.2	100
Lower storeys	1.2–3	1.5–3	10
Shelter partly above ground			
with 60-cm earth cover	12–50	15–30	50–200
with 90-cm earth cover	20–100	50–150	200–1 000
Concrete blockhouse shelter			
with 20-cm walls	2–3	5–10	10–150
with 30-cm walls	2.5–5	10–20	30–1 000
with 60-cm walls	5–10	50–150	500–10 000

As is seen, light-frame houses offer little protection against initial radiation. Basements offer more protection, but in some countries the majority of houses have no basements. Upper stories of residential blocks offer much more protection from fall-out gamma-rays than do the lower stories. A high protection is offered by a concrete blockhouse shelter with walls 60 cm or more thick.

6.5. *Shelters*

Most custom-built shelters are designed as blast shelters rather than for protection against fire-storms or radiation, that is, they are built to resist overpressures of 100 kPa or higher.

As was mentioned earlier (in section 2.2.2), one possible thermal effect is a fire-storm which might rage over an area much larger than that covered by the direct heat wave. For example, with a 1-Mt bomb the area covered by the fire-storm may be 500 km^2. Within that area, even in shelters specially designed for protection against the other effects of the bombing, the occupants may perish through asphyxiation, the fire-storm having drawn out the oxygen from the shelter, or they may be cremated in the very high temperatures.

If the detonation occurs sufficiently far from the shelter for the persons in it to survive the blast wave and the fire-storm, the shelter would also provide adequate protection from the initial nuclear radiation. But the early fall-out creates problems which make the design of shelters much more complex and costly; in the case of an all-out nuclear war these problems are likely to be unsolvable.

The main problem relates to the length of stay in the shelters. In a conventional war one can leave the shelter as soon as the bombing is over. This is not the case with nuclear bombs which produce local fall-out. It may be necessary to remain in the shelter for weeks and, unless radiation monitoring equipment is available in the shelter and is in working condition despite the electromagnetic pulse, the occupants will not even know when it is safe to come out.

Long occupancy carries with it problems of the provision of clean air, food, water, heating or cooling. It is very likely that electrical power stations will be primary targets in a large-scale war; the supply of electricity will thus probably be disrupted and this will in turn affect the supply of water. Emergency power supplies can cope with the needs only for a short time. Under these conditions the problems of staying in shelters will gradually become insurmountable, and they will force the occupants to leave the shelter even if the level of radiation outside is still dangerously high.

Inability to control the temperature and humidity of the air will cause much discomfort and will affect health. Sanitation will create enormous difficulties if prolonged sojourn in shelters is necessary. Absence of adequate supplies of water will make it impossible to maintain even elementary sanitary facilities. Many of these also depend on the supply of electricity. Uncollected refuse and faecal material will encourage disease, particularly since bacteria are hardly affected by radiation.

Those who arrive at the shelter after some delay and who have received a high dose of radiation, or persons in the shelters who have been outside on some essential mission—as is bound to happen—will begin to show symptoms of radiation sickness. Infections may soon begin to affect the occupants, particularly the children who nowadays have no immunity to many infectious diseases.

Added to all these problems will be the psychological effects: anxiety about the fate of missing relatives; distress in observing symptoms of sickness in others, particularly the very unpleasant symptoms of radiation sickness; traumatic effects of witnessing agony and death; and emotional upsets at the sight and smell of decomposing bodies.

In combination with the physical discomforts of life in shelters, this will lead to social stresses and to aggressive or violent behaviour.

Taking all this into account, it appears that the large protection factors provided by the great thicknesses of earth or concrete will be illusory in the final outcome.

6.6. Evacuation of populations

In the absence of shelters for the whole population, or even when shelters are available but many people do not wish, for the reasons stated above, to be cooped up in them, the alternative form of civil defence is the evacuation of the population from areas where fall-out is expected, perhaps even before the war has started. An immense organizational machinery will have to be involved and many logistic problems solved in carrying out such an operation, including the inevitable traffic jams in a panic exodus, particularly if the evacuation of large cities is involved. But even assuming that this evacuation machinery will still be operational after a nuclear attack, the main problem will be where to evacuate to. In a conventional war evacuation is usually planned into rural areas, where military activities are less likely to take place than in urban areas. But with a nuclear war, the land covered by fall-out could involve an area of many thousands of square kilometres for each detonation and, as was shown in section 4.2.3, the cigar-shaped area could stretch in any direction depending on the meteorological conditions at the time of the explosion. Thus, if evacuation takes place before the bombs are dropped, the population might actually be evacuated *into* an area of high fall-out. Even after an attack has taken place, it will be impossible to predict accurately where the fall-out will occur. If the evacuation plan is initiated after the bombs are dropped, when the direction of fall-out becomes known, then the whole purpose of the evacuation might already have been forfeited. As was shown earlier, the intensity of fall-out decreases rapidly with time and the largest dose is received soon after the fall-out particles descend. By the time the evacuation has been organized and accomplished it might well be that the evacuation centres will be reached mainly by sick and dying people; this will be likely to spread disease and panic in areas not affected by fall-out.

All this assumes that there will still be areas free from fall-out after the bombing. For small countries, like some in Europe, this might not be the case even in a limited nuclear war.

6.7. Medical aspects of civil defence

Whatever the type of civil defence measures taken, there are likely to be many people with injuries requiring medical attention. Indeed, this problem will be much more agonizing with civil defence measures than without, because with civil defence many people are likely to survive who would otherwise have died immediately. But in the absence of medical care the majority of these survivors will die later, often very painfully.

Most doctors, nurses and hospitals are situated in cities. If these are targets for nuclear bombs, the majority of the medical personnel will be killed and hospitals destroyed. Many drugs in undamaged stores will become useless in the absence of refrigeration. If fall-out conditions necessitate staying indoors or in shelters, the surviving doctors will be unable to go to the injured in the initial period, when most of the wounds and burns will have occurred. Even after that period it will be physically impossible for the surviving doctors and nurses to cope with but a small fraction of those needing attention.

In a study made in 1962 of the aftermath of a nuclear war which could have included a 28-Mt attack on the Boston metropolitan area, it was concluded that three-quarters of all physicians in the area would be killed immediately or fatally injured, and that only about one-seventh would be in a condition to render post-attack medical care. There would be about 1 700 injured persons to each of the functioning doctors. If each of these doctors spent an average of 10 minutes with each injured person and worked 20 hours each day, it would take 14 days before each casualty could be seen for the first time. This estimate was thought to be rather optimistic [27].

A more recent assessment of the problem was made by a group of doctors and medical scientists from 14 countries at a Pugwash Conference [28]. They felt bound to issue the following warning.

1. MEDICAL DISASTER—PLANNING FOR A NUCLEAR WAR IS FUTILE.

A nuclear war would result in human death, injury, and disease on a scale that has no precedent in history, dwarfing all previous plagues and wars. There is no possible effective medical response after a nuclear attack—in one major city alone, in addition to the hundreds of thousands of sudden deaths, there would be hundreds of thousands of people with severe burns, trauma, and radiation sickness—all demanding intensive care. Even if all medical resources were intact, the care of these immediate survivors would be next to impossible. In fact, most hospitals would be destroyed, medical personnel among the dead and injured, most transportation, communication and energy systems inoperable, and most medical supplies unavailable. As a result, most of those requiring medical attention would die.

Medical problems that would be minor in normal times—infections and fractures for example—would prove fatal for many. Numerous deaths would also occur from the interaction of multiple, simultaneous injuries which would be trivial if each occurred singly. Large numbers of those who escaped an acute death would suffer mutilating injuries. Furthermore, under the conditions of rampant chaos and terror, the incidence of psychiatric disorders would sharply rise. The risk of long term effects, such as cancer, would increase during their entire lifetime for many survivors, and possibly for their offspring as well.

2. EFFECTIVE CIVIL DEFENCE AGAINST A NUCLEAR ATTACK IS IMPOSSIBLE.

Bomb shelters in cities under nuclear attack would be useless owing to the blast, heat, and radiation effects. Shelters as far as ten kilometres from the centre of even a one megaton surface nuclear explosion could become ovens for their occupants—the great surface fires would cook and asphyxiate them. At greater distances, shelters would provide only temporary protection against the high levels of radioactive fall-out. In a nuclear war, one would emerge from a shelter into an environment that was a nightmare—water would be undrinkable, food contaminated, and the economic, ecologic, and social fabric, on which human life depends, destroyed. For the survivors, the risk of epidemics would be great, as a result of: the unburied human and animal corpses everywhere; multiplication of viruses, bacteria, fungi, and insects, which are highly resistant to radiation; and the high sensitivity to radiation of the human body's ability to fight infection.

In sum, there are no defences against the lethal effects of nuclear weapons, and there is no effective treatment for those who initially survived a nuclear attack. Under all conditions, medically, nuclear war would be an unparalleled catastrophe.

As doctors of medicine and scientists in health-related fields, we conclude, therefore, that nuclear weapons are so destructive to human health and life that they must never be used. Prevention of nuclear war offers the only possibility for protecting people from its medical consequences. There is no alternative.

7. Other warlike uses of radiation

In a world in which every conceivable lethal agent has been considered as a means of warfare and in which many are actually deployed, it is not surprising that radiation, with its peculiar and all-pervading characteristics, should be considered as a tool of destruction. There is a distinct possibility that ionizing radiation other than in the form of nuclear weapons may be put to military or terrorist use. Although an analysis of the military effectiveness of radiation shows that—except in certain situations—it has hardly any real advantage over existing types of weapon, it is nevertheless an addition to the instruments of war, a lethal agent to which modern society is highly vulnerable. As such it might be employed in some circumstances.

7.1. Attacks on civilian nuclear power installations

In any future war, electricity-generating power stations are likely to be primary targets of attack, because their destruction could paralyse the whole war effort of a country. The trend towards ever larger power stations, arising from economy of scale, makes each such station a highly attractive target. These installations will no doubt be strongly defended, but the greatly improved performance of modern missiles—in terms of (even non-nuclear) explosive power, range, payload and accuracy—ensures a high degree of success in such attacks.

All this applies especially to nuclear power stations which contain reactors of very high output—sometimes two or more in one station—and which may provide a significant proportion of the energy needs of a country. Putting them out of action could have a devastating effect on the economy, particularly in countries which plan to obtain most of their electricity from nuclear reactors, quite apart from the huge material loss, since reactors represent a very large concentration of capital investment.

There may be another important reason for making a nuclear power station a primary target, namely the release into the biosphere of an immense quantity of radioactivity. Such a release, with the consequent contamination of a large area and the panic that this would cause among the population, might indeed be the main purpose of the attack.

7.1.1. Attack on nuclear reactors with conventional weapons

In the course of its operation, a nuclear power reactor accumulates a vast amount of radioactivity in its fuel elements. This radioactivity is produced during the fission of uranium or plutonium. It is therefore of the same origin as that produced by fission weapons, but the composition is considerably different. One reason is that the frequency of occurrence of the initial fission products is different in a thermal reactor, where fission is caused by slow neutrons, from that in a bomb, where the chain reaction is propagated by fast neutrons (see section 4.2.1.1). What is more important is that in a nuclear bomb only the primary fission fragments are present at the instant of the explosion; most of these are short-lived nuclides which subsequently decay into nuclides of longer half-life. In a nuclear reactor these longer-lived nuclei are steadily built up, being continuously produced by the decay of the short-lived nuclides. As a result, the proportion of fission products of long half-life is much greater in a reactor than in a bomb. The overall decay of the radioactivity is therefore much slower, and its time dependence is different from that in fallout, as was described in section 4.2.2. Figure 29 shows a comparison of the decay of radioactivity from a 1-Mt bomb with that from a 1 GW(e) light water reactor after several years of operation, with the fuel elements being replaced every three years. The decay of the activity in the reactor is reckoned from the moment of its shut-down. As is seen, the total activity in the reactor is initially about 100 times lower than that from the bomb, but they become equal after about four days. Thereafter, the reactor activity exceeds that in the bomb and the difference becomes increasingly larger; for example, after one year the reactor activity is still as intense as the bomb activity was after one month. Thus, radiation doses resulting from a reactor release could remain at unacceptably high levels for a much longer time than those from a bomb which initially produced the same activity.

It should be noted that the curves in figure 29 refer to the *total* radioactive content of the bomb or reactor, at the times shown. The proportion of the reactor radioactivity released into the atmosphere, and therefore the extent of the radiological hazard that may result from an attack on a reactor, would depend on the structure of the reactor and the means used to destroy it. Reactors are usually contained in vessels with walls, base and ceiling of reinforced concrete several metres thick; this is the biological shield needed to reduce the intensity of radiation outside the vessel to an accepted level. Additionally, in many reactors the core, that is, the assembly of the fuel elements, is contained in a pressure vessel designed to withstand the high pressure of the water used as a neutron moderator and as a coolant. But despite this heavy protection, modern precision-guided bombardment with conventional weapons could succeed in rupturing the containment vessel as well as the pressure vessel. Alternatively, the task might be achieved in a commando raid, as was carried out on a heavy water plant during World War II.

Should both the containment and pressure vessels be breached by the use of explosives, a proportion of the fission products, as well as the plutonium and the other actinides, would be released and dispersed by the force of the explosion over a considerable area. But since the explosive power used is likely to be less than 1 kt of TNT, and the temperature increase would be much lower than

Figure 29. Decay of activity from: (*a*) a 1-Mt bomb, and (*b*) a 1-GW(e) reactor

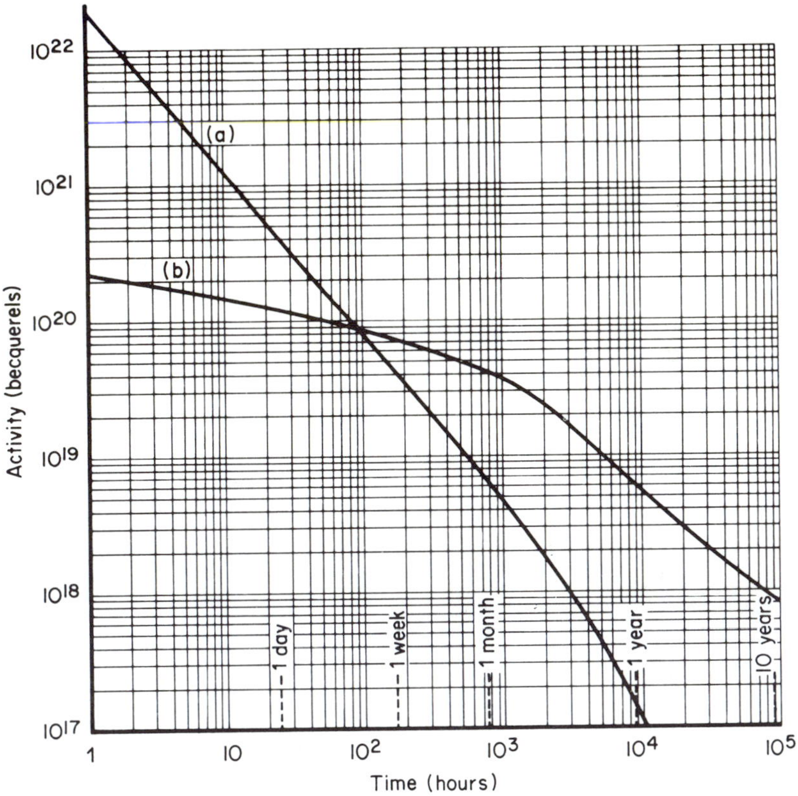

in a nuclear explosion, the radioactive plume would not rise to great heights, and the area of contamination would be much smaller than the area of local fall-out produced by a nuclear bomb, although over the area where it is dispersed the intensity of the radioactivity from the reactor could be very high. However, very serious consequences could ensue even if the pressure vessel were not ruptured but—as is very likely—the cooling system of the reactor were put out of action. This would lead to a core melt-down. The heat produced by the absorption of the radiation emitted in the process of radioactive decay—which goes on relentlessly even after reactor shut-down—is so intense that, unless dissipated by cooling, it would raise the temperature sufficiently to melt the reactor core and the containment vessel; the hot mass would burn its way through the foundation and an immense amount of radioactivity, equal to that from thousands of tons of radium, would stream out. In a pressurized water reactor the melt-down of the core could occur within less than one minute after the loss of coolant; with other types of reactor it might take a few minutes. This sequence of events is popularly known as the

"China Syndrome". Other likely events following the loss of coolant and the melting of the core are a steam explosion, if the molten core gets into contact with water, or the generation of hydrogen which—after combining with oxygen—would cause an explosion. In both these cases, the containment vessel may be breached and the radioactive materials scattered into the atmosphere.

These possibilities, as well as a number of other events, were considered in studies of the consequences of accidents that could occur in a reactor in the course of routine operation. Calculations have been made of the number of casualties and the extent of damage that might result from different types of accident. One result of such calculations is shown in table 26 which lists the consequences of what is described as an "extremely serious accident". In peace-time such an accident to a reactor would be considered an extremely rare event, but the consequences listed in the table may follow a deliberate military attack on a reactor.

Table 26. Consequences of an extremely serious reactor accident

Consequence	Extent	
Prompt fatalities	3 300 persons	
Early illness	45 000 persons	
Thyroid nodules	240 000 persons	(spread over 30 years)
Latent cancer fatalities	45 000 persons	(spread over 30 years)
Genetic defects	30 000 persons	(spread over 30 years)
Economic loss due to contamination	$14 \times 10^9	
Contamination area	8 000 km²	

Source: Keeny, S.M., *Nuclear Power Issues and Choices,* Ford/Mitre Report (Ballinger, Cambridge, Mass., 1977), p. 224.

Appalling as this scale of devastation appears, it fades into insignificance when compared with the outcome of a nuclear war. On the other hand, there is a greater probability of its occurrence, since it may be brought about without a nuclear war, and conventional wars are only too frequent. We must thus conclude that radiation could become a significant agent of death and destruction in a world increasingly populated with nuclear reactors.

7.1.2. Attack on nuclear reactors with nuclear weapons

In a nuclear war, should nuclear reactors be attacked with *nuclear* weapons, the consequences could be very serious, making civil defence measures even less effective and adversely influencing the prospects of post-war recovery. This follows from the much slower decay of the radioactivity from reactors, as described in the previous section.

An attack on a reactor with a nuclear weapon, even of relatively low yield, say 100 kt, would vaporize the reactor, and much of its radioactive content would be sucked up with the fire-ball, carried with the wind and subsequently deposited as local fall-out, together with the fission products of the bomb. The area covered by the fall-out will be approximately the same as from the detonation of the bomb alone, but the dose rate of the deposited activity will be a different function of time, and will depend both on the explosive yield of the bomb and on the radioactive content of the reactor. For example, if the bomb were 1-Mt, and the capacity of the reactor 1 GW(e), the fall-out dose would initially be mainly from the bomb, but after a few days the dose from the reactor activity would become greater and would thereafter predominate.

The effect of this on the possibility of return to normal occupancy of an area which had been evacuated before the bomb, has been recently studied [29]. Some of the findings are depicted in figure 30, which shows the contours of the fall-out areas in which a dose of 1 Gy could be accumulated in one year, starting from one month after the explosion; contour (*a*) gives the area for the detonation of a nuclear bomb alone, and contour (*b*) for an explosion of the same bomb on a reactor. A surface explosion of a 1-Mt bomb was assumed. The reactor was assumed to have a capacity of 1 GW(e) and the activity released was about one-fifth of the total content.

Figure 30. Contours of 1 Gy fall-out dose in one year, starting one month after the detonation of (*a*) a 1-Mt bomb, and (*b*) a 1-Mt bomb on a 1-GW(e) nuclear reactor

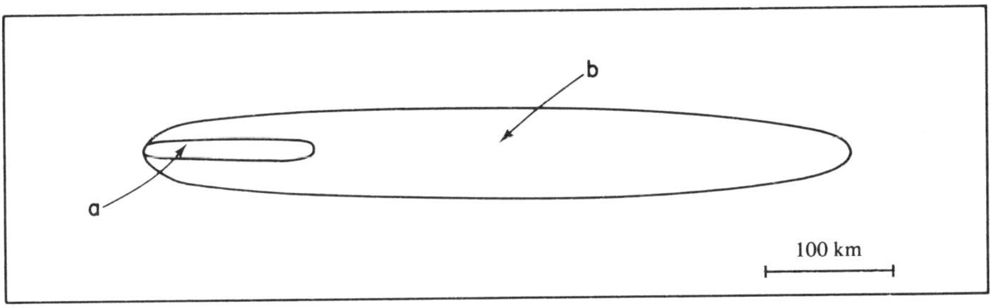

Table 27 shows areas for several values of accumulated dose for the same periods as in figure 30. As is seen, the reactor activity would make huge areas uninhabitable for a long time. In Western Europe, where many reactors are in operation and many more are being built or planned, an attack on them would make practically the whole of these and neighbouring countries uninhabitable by ordinary standards of radiation safety. Such an attack would seriously upset planned civil defence measures either by evacuation or by shelters.

Table 27. Areas affected by detonation of nuclear weapons alone and on nuclear power facilities

Dose[a] (Gy)	Area (km^2) 1-Mt bomb	1-Mt bomb on 1 GW(e) reactor	1-Mt bomb on a storage tank
1	2 000	34 000	61 000
0.5	4 000	46 000	87 000
0.1	25 000	122 000	164 000

[a] Dose accumulated in one year starting one month after detonation.

7.1.3. Attack on nuclear reprocessing facilities and storage tanks

Other installations associated with peaceful nuclear power which contain large quantities of radioactivity are cooling ponds, reprocessing plants and storage tanks of spent nuclear fuels. Cooling ponds, usually situated near reactor sites, contain the fuel elements removed from the reactor before they are shipped to reprocessing plants. Storage tanks contain 'high-level' radioactive wastes, that is, the fission products after they have been through the chemical separation process in the reprocessing plant. In both cases cooling is essential for a long time to dissipate the heat from the radioactive decay. Deliberate or incidental disruption of the cooling systems, or of the containment vessels, would result in the release of large amounts of radioactivity. Reprocessing plants, where the spent fuel elements are chemically treated to separate the uranium and plutonium from the waste products would be particularly vulnerable. Such plants have large capacities and can handle the outputs from many reactors. The destruction of a reprocessing plant or of a storage tank may result in the release into the atmosphere and the aquatic biosphere of amounts of radioactivity greatly exceeding those from a reactor core melt-down.

Should a nuclear weapon be exploded on a reprocessing plant or a storage tank, the uninhabitable areas would be even greater than following an attack on a reactor; they would remain uninhabitable for a longer period because the activity in these plants or tanks is even longer-lived than that in a reactor. As an example, the last column of tablc 27 shows the areas covered by doses received in one year, starting one month after the explosion of a 1-Mt bomb on a storage tank of a capacity of about 3 000 cubic metres.

There are as yet few reprocessing plants and storage tanks, and it is less likely that these installations would be prime targets in a conventional war since their destruction would not have the same immediate crippling effect on industry as the incapacitation of a large power station. But circumstances may occur, in a non-nuclear war, when an attacker would want to exploit the panic which would inevitably follow the news of a large release of radioactivity, or when the attacker may want to make recovery in the period after a nuclear war more difficult.

7.2. Radiological warfare

The possibility of using radioactivity as a means of warfare has been discussed for many years. Although there is no evidence that such use has been contemplated by any country, or that any preparation for radiological warfare is being made anywhere, the concern about it was sufficiently strong for the two great powers to submit in 1979 a joint proposal to the Committee on Disarmament for a treaty prohibiting the development, production, stockpiling and use of radiological weapons [30]. In this proposal radiological weapons are defined in the following terms:

(1) Any device, including any weapon or equipment, other than a nuclear explosive device, specifically designed to employ radioactive material by disseminating it to cause destruction, damage or injury by means of the radiation produced by the decay of such material.

(2) Any radioactive material other than that produced by a nuclear explosive device, specifically designed for employment, by its dissemination, to cause destruction, damage or injury by means of the radiation produced by the decay of such material.

The submission of this Soviet–American proposal could be interpreted as a genuine concern by the great powers that radiological weapons may become an important means of warfare and as a fervent desire that such development be prevented. However, it could also have another interpretation: the great powers know that radiological weapons do not have much military value and that because of this there would be no serious obstacle to the ratification of the proposed treaty. But the very fact of agreeing to such a treaty would make a good impression, as if something positive had at long last been achieved in the field of disarmament. This cynical interpretation gains in credibility when one considers the careful wording of the definition of radiological weapons, which excludes nuclear explosive devices. Thus, the neutron bomb, which is specifically designed to kill by radiation, would be allowed by the terms of the treaty, since the radiation would arise as a consequence of a nuclear explosion. An attack on a nuclear reactor would also be allowed if it were carried out with a nuclear weapon.

Several situations can be envisaged in which radiological weapons might be considered militarily expedient; the purpose might be either defensive or offensive. Examples of defensive use are: to make an area impassable for enemy troups; to halt an armoured attack; to inhibit a river crossing; or to close a mountain pass.

Offensive uses are more plausible. The objective might be to force the evacuation of populations from cities or other industrial or communication centres, thus playing havoc with the economy and fighting potential of the country attacked; or it may have a more sinister purpose—to kill a large number of people. These objectives could of course be achieved more effectively by nuclear explosives, but radiological weapons may be preferable if the attacker intends to occupy the country and would prefer to avoid the structural damage caused by nuclear bombs. A more likely reason is that the attacker might not have nuclear weapons, or does not want to start a nuclear war.

Radiological weapons might be employed by making bombs or shells packed with radioactive materials and delivering them by means of ordinary (non-nuclear) explosives. Alternatively, rockets or cruise missiles might be used. The radioactive materials could also be disseminated in the form of liquid or solid particles by aerial spraying. For defence purposes, strips of radiological agents could be laid using ground or airborne carriers.

The radioactive materials for radiological weapons could be fission products from reactors; or plutonium and other actinides, also from reactors;

or artificially produced radioactive nuclides. But it is shown below that none of these make militarily effective weapons of war.

7.2.1. The use of fission products

Present concern about the possible employment of radiological weapons stems mainly from the increasing use of nuclear energy for peaceful purposes and the resulting production of immense quantities of radioactive materials. A typical reactor of 1 GW(e) capacity contains—after three years of operation—nearly 3 tons of 'waste products', most of which are radioactive. With the nuclear power capacity in the world, some 145 GW(e) at the end of 1980, possibly rising to 1 000 GW(e) by the year 2000, the total radioactive content of reactors would be truly colossal. Moreover, should nuclear power become widespread, the radioactive waste products would be available to a large number of countries. Thus, in theory radiological warfare appears to present a serious danger. In practice, however, the employment of fission product radioactivity for war purposes is of limited use and questionable military value.

One of the reasons relates to the difficulties in the fabrication of the radioactive materials in a form suitable for a weapon. Unlike fission bombs, where all the radioactivity is created at the moment of the explosion, in a radiological weapon the fission products—after reprocessing—would have to be assembled beforehand and packed into missiles or into aircraft for spraying. All stages of preparation would present very serious problems of protecting the personnel from over-exposure to radiation.

The waste products from reactors are usually mixed with large quantities of extraneous materials. To obtain a high concentration of the radioactivity, chemical separation, not normally envisaged in a civilian nuclear power programme, would have to be carried out. For spraying purposes, the separated waste products would have to be fabricated into aerosols or into liquids. All this means that the mere possession of radioactive materials in the form of reactor waste products is not sufficient, and that considerable expenditure and technological know-how would be required to convert the waste products into a radiological weapon.

However, the main limitation to the use of fission products from a reactor would result from the fact that a relatively high proportion of them have long half-lives. This fact would rule out the use of waste products for defensive purposes, because it would mean making one's own territory uninhabitable for many years. But this would also make offensive use impracticable. The best way to create a radiation level high enough to deliver lethal doses or to force evacuation of populations is by using short-lived nuclides. One reason for this is that the activity per unit mass is very high for short-lived nuclides; this means that even a small mass of material would be sufficient to cover a large area with a deadly blanket of radiation. Another reason is that the radioactivity would have disappeared after a short time, enabling the attacker to enter the area or occupy the country. But, as figure 29 shows, unlike the radioactivity from the bomb, the decrease in activity of reactor material is quite slow. It would probably take at least one week from the time of removing the fuel elements from the reactor to fabricate the waste products into a suitable form for dissemination. During the period from one to six weeks the activity decreases only by a factor of 2; during the same period the bomb activity decreases by nearly a factor of 10. Between 6 weeks and 10 years the reactor activity decreases by a factor of 30; if, therefore, a sufficiently large amount is deposited to produce a lethal level of radiation during the first few weeks, the area will remain dangerously contaminated for years. Moreover, the radioactivity may be carried to neighbouring countries, perhaps even into the territory of the attacker.

A spread of activity occurs also from fall-out in a nuclear war, but—as discussed in chapter 5—such a war is in any case likely to be suicidal for those who wage it.

7.2.2. The use of plutonium and other actinides

The civilian nuclear power programme also gives rise to the production of huge quantities of plutonium and other alpha-emitting radioactive substances, mainly the actinides. If the total nuclear power capacity in the world were to grow to 1 000 GW(e) by the year 2000, the total accumulated amount of plutonium could be about 2 500 tons. If such an amount were to be produced by nuclear weapons it would require an explosive yield of 100 000 Mt (see section 4.2.1.3). Great care would of course be taken to prevent the diversion of the plutonium for illicit purposes but, despite the safeguards system set up by the International Atomic Energy Agency, it is likely that considerable quantities of plutonium would be accessible to all nations with a nuclear power programme.

A significant advantage of plutonium as a radiological weapon is that since it decays by the emission of alpha-particles, which are very easily stopped, the problem of radiation shielding is greatly reduced and it would thus be much easier for a nation (or a subnational group) to fabricate the plutonium in a form suitable for dissemination. It should also be noted (see section 3.7) that plutonium is considered to be one of the most radio-toxic substances; the deposition of less than 20 mg in the lung could cause death within one month, and much smaller quantities could result in delayed but nevertheless fatal cancers. The psychological effect of contaminating a populated area with plutonium would be very large. Plutonium is thus potentially a very powerful lethal weapon, although in practice its use would be very limited.

For use as a radiological weapon, plutonium would probably be dispersed in the form of an aerosol of finely divided particles. In open areas it would be impossible to maintain the suspension for any length of time. Even a gentle breeze would rapidly disperse it over large areas, and a very high initial concentration would thus be necessary. This concentration would have to be sufficiently high so that a lethal dose would be inhaled in a short time, say one hour. If 20 mg is taken as the lethal acute dose, the necessary concentration, assuming little dispersion, is about 30 mg of plutonium per cubic metre. To cover an area of a medium-sized town, say 100 square kilometres, to a height of 3 metres would require about 10 tons of plutonium. This rules out plutonium as an acute lethal weapon. However, quantities smaller by a factor of 1 000, that is, a few kilograms, would be sufficient to require the evacuation of the population and necessitate difficult decontamination measures.

But the main limitation would arise from the very long half-life (24 100 years) of the chief isotope of plutonium. Once it was disseminated its harmful action would continue practically for ever and would gradually spread through atmospheric or aquatic vectors to other parts of the world.

Of much shorter half-life (86 years) is another isotope of plutonium, ^{238}Pu. It is being produced in large quantities for various purposes, such as a source of power in satellites or in pacemakers. Because of the shorter half-life of ^{238}Pu, quantities nearly 300 times less than those of ^{239}Pu would be needed to produce the same radiological effects. There are also other actinides with much shorter half-lives than ^{239}Pu, but they are still too long to make them of military interest.

7.2.3. The use of short-lived radioactive nuclides

As was shown above, there is little military rationale for long-term radioactive contamination. From the military point of view the most advantageous would be radioactive nuclides of short half-life. Nuclides such as potassium-38 or calcium-49, with half-lives of a few minutes, could deliver a very high dose in a short time and for all practical purposes would have disappeared after an hour. Other nuclides, such as sodium-24, silicon-31, potassium-42 or manganese-56, have half-lives of the order of hours; in this case about a week would have to elapse before the area over which they were deposited could be considered free from contamination.

The problems with these materials lie with their production and delivery. They cannot be stored and have to be made just before delivery. Short-lived radioactive nuclides can be made only by bombardment of non-radioactive materials with intense beams of charged particles in accelerators, or with neutrons in reactors. The bombardment would have to be of short duration, because the longer it lasts the greater is the production of other long-lived nuclides. At the present time only reactors contain a sufficiently high neutron intensity to make this possible, but some modifications to their structure would probably be necessary to enable large quantities of artificial radioactivity to be produced. Even so it is doubtful whether quantities of military interest could be manufactured. For the nuclides mentioned above the specific activity (activity per unit mass of material bombarded) would be thousands of times smaller than that of the fission products. A greater specific activity can be achieved for some nuclides by chemical separation, but the short half-life restricts the use of this technique. The low activities attainable would make the artificially produced radioactive materials a very ineffective radiological weapon.

The additional difficulty with short-lived nuclides is that much of the activity would have disappeared between the time of production and delivery. This would certainly rule out nuclides with half-lives of the order of minutes.

Furthermore, since all the nuclides of interest are gamma-emitters there would also be problems in protecting personnel from radiation.

We must thus conclude that, for different reasons, the use of radioactive materials as a tool of war is unlikely to be of military significance, although it could be of great nuisance value.

7.3. Terrorist activities

Although terrorist activities are not usually classified as warfare, the increasing financial and technological resources becoming available to terrorist groups may raise the level of damage resulting from terrorist actions to approach that from a small war. In any case, the definition of terrorist activities is vague and may include those of any national sub-group or guerrilla movement fighting against a specific government or system of government. The use of nuclear weapons or nuclear materials by terrorists has so far occurred only in fiction, but considering the rapid increase in nuclear weapon capabilities and nuclear power installations it is only a matter of time before they occur in reality.

One characteristic feature which supposedly distinguishes terrorist activities from the military acts of nations is that terrorists usually resort to blackmail and threaten to perpetrate heinous deeds if their demands are not met. But this distinction is hardly valid when it is considered that the whole concept of deterrence in a nuclear war is in fact one of blackmail, with each side holding the other to ransom. The element of unpredictability, which is generally associated with terrorist activities, is also part and parcel of government policies in relation to the use of nuclear weapons, each side deliberately creating uncertainties about its response to a specific crisis.

One of the alleged motivations of terrorist activities—desperation—can also play an important role in inter-nation relations; the use of nuclear weapons by a nation as an act of desperation is one of the distinct possibilities in a nuclear world. It is in fact difficult to find a qualitative difference between the motivation for hostile actions of sub-national and those of national groups. As far as nuclear weapons are concerned, the main difference is a quantitative one, that is, whether or not a sub-national group would be able to muster the effort needed to present a credible menace.

While the number of persons involved in non-nuclear terrorist actions may range from one individual to a significant proportion of the population, in the case of nuclear terrorism it has always been assumed that a suitably large group with considerable financial and technical resources would be needed as a minimum. This may be true in the current situation, but in the course of time, if nuclear weapons spread to many countries, and if plutonium becomes a staple commodity as part of a large nuclear power programme, it will be progressively easier for small groups to acquire the means of nuclear blackmail.

Terrorist activities in the nuclear field could take different forms. For example, they might gain possession of nuclear weapons, take over a reactor or reprocessing plant, or acquire radioactive waste products or plutonium.

The most dangerous terrorist act would be a claim of having acquired a nuclear weapon, coupled with a threat to detonate it in a large centre of population. How seriously a government would take such a threat would depend on the credibility attached to the terrorists' claim and the assessment of their capability to achieve a high explosive yield if the bomb were detonated. If it were known to the government—even if not to the public—that one or more bombs had been stolen from the nuclear arsenals, the threat would have to be taken extremely seriously. A more likely way for terrorists to obtain a nuclear weapon, however, would be to acquire a sufficient quantity of plutonium and manufacture the bomb themselves. As the world nuclear power capacity increases and large quantities of plutonium are processed and transported, there is an increasing likelihood that an amount of plutonium sufficient for one or more bombs could be diverted in such a way that the diversion remained undetected for some time. In these circumstances it would be very difficult for a government to ignore a claim by a terrorist group that they were in possession of a sufficient amount of plutonium.

Once the material was acquired the actual production of the bomb would not present too much difficulty for a group with adequate resources. Various estimates have been made about the magnitude of the necessary effort. According to one estimate a few persons, possibly even one, with possession of about 10 kg of plutonium oxide and a substantial amount of chemical high explosives, could design and build a crude fission bomb with a yield of at least 100 tons, within several weeks [31]. However, another estimate is that, as a minimum, a team of more than a dozen well-trained persons would be needed, and that the whole project would take about two years at a cost of a few tens of millions of dollars [32]. Even this is well within the means of some terrorist groups operating today. Moreover, this estimate was made a few years ago, and it is likely that as time goes on and more nations acquire nuclear weapons or the potential to make them the effort needed by a terrorist group will become much smaller.

Even if the claim by the terrorists that they had a nuclear bomb were not contested, there would still be uncertainty about the yield which they could achieve when detonating a bomb with reactor-grade plutonium. Such plutonium has a considerable admixture (about 25 per cent) of plutonium-240, a nuclide with a high probability of undergoing fission spontaneously with the emission of neutrons; one kilogram of ^{240}Pu emits about a million neutrons per second. Thus, with reactor-grade plutonium there is a significant probability that a neutron will appear and initiate the chain reaction prematurely (see section 2.1.1), resulting in a pre-detonation and a greatly reduced explosive yield. Actually, with a high-speed implosion assembly the probability of pre-detonation is quite small, but even with a less sophisticated technology the explosive yield is statistically distributed round a lower mean, and an explosive power of some kilotons may be achieved even at the low end of the distribution. In any case, it has been reported that a test of a nuclear weapon using reactor-grade plutonium has been carried out and gave a high yield. (It should be noted that the critical mass of reactor-grade plutonium is about 50 per cent greater than the values given in section 2.1.1). In view of this, it seems unlikely that governments would gamble on the chance that the yield would be low, and defy the terrorists.

Evacuation of populations would be impractical if the terrorists chose not to tell in which city the bomb was planted and when it was timed to go off. Under these circumstances it is highly probable that governments would give in, unless the terms dictated by the terrorists were completely unacceptable.

Another possibility is of a terrorist group taking over a nuclear reactor or a reprocessing plant and threatening to blow it up. Such an act would have to be carried out like a military operation, requiring a well trained team armed with modern weaponry. Specialized training in nuclear matters would not be essential, but knowledge of the mode of operation of a reactor or reprocessing plant would mean a greater likelihood of causing serious damage and would increase the credibility of the attackers. If a group took over a reactor they would not need to blow up the heavy biological shield of the pressure vessel; all they would have to do would be to cut off the supply of cooling water to bring about core melt-down. The possible extent of the resulting damage was described in section 7.1.1.

Somewhat less frightening, but still very serious and easier to bring about, would be a threat by terrorists to use radioactive materials as a radiological weapon. The waste products after reprocessing the spent nuclear fuel elements from reactors do not come under the IAEA safeguards, and a group could get hold of the waste products relatively easily. There would of course be considerable difficulty in handling intensely radioactive materials, and the terrorists would have to be equipped with special tools and containers. They would also have to devise credible means of carrying out their threat, but all these problems are less difficult to overcome than to assemble a nuclear bomb. For example, if ordinary explosives were placed inside a mass of waste products and then detonated by remote control, the radioactivity could be dispersed over a huge area, which would become uninhabitable for a long time; a credible enough threat.

The fact that the radioactive materials have long half-lives—which, as was shown earlier, makes them unsuitable as a tool of war—would hardly be a deterrent for terrorists. On the contrary, the longer-lasting the effect of the radioactive contamination, the more serious the threat it would present to the community, and this would strengthen the hand of the terrorists. In this case too, evacuation of the population would not be a feasible response, if the time and place of carrying out the threat were not revealed by the terrorists. The undisclosed target could be of many types; it might be a harbour which could be put out of use for many years; it might be a large water reservoir which could be poisoned; or it might be a vital industrial or communication centre which could be made inoperative for a long time. Although the number of deaths from acute radiation exposure would be relatively small, the long-term effect could have very serious economic consequences. The dissemination of plutonium inside large buildings, in quantities smaller than those needed for a nuclear weapon, could also pose a serious threat.

In summary, while as a military tool radioactivity does not appear to be effective and is therefore unlikely to be used in warfare, its employment by terrorist groups is a distinct possibility, and the ever-existing threat of its use may make normal life increasingly difficult.

The fact that there are many other poisons and contaminants which terrorists could use to achieve their aim does not mitigate the danger of radiation. At the least it adds to the other dangers of modern life, but radiation is more likely to be employed as a tool of terror because of the psychological effect which a potential release of radioactivity would have on the public. This singular and somewhat irrational fear is, however, fully justified when one remembers the terrible short-term and long-term consequences of the use of radiation in nuclear warfare.

8. Conclusion

Of the three main injurious agents of nuclear weapons—blast, heat and ionizing radiations—one aspect of the last agent, fall-out, is the least amenable to quantitative assessment, owing to its dependence on a number of unpredictable factors. Yet, in any nuclear war, local and/or global fall-out are likely to produce a heavy casualty toll.

In a counterforce attack, with the underground ICBM silos as targets, the great accuracy of modern missiles might make it possible to avoid direct hits on large centres of population, but millions of civilians are likely to be killed or suffer long-term effects of radiation from exposure to local fall-out. In an all-out war, into which any nuclear conflict is very likely to escalate, the largest immediate casualty toll would be from the effects of blast, heat and initial radiation in the cities hit by nuclear weapons. But in this case too fall-out would add immensely to the numbers of dead and injured, as well as greatly diminishing the extent of post-war recovery. Civil defence measures such as deep shelters, which may provide some protection against blast, would be much less effective against local fall-out, owing to problems arising from the necessary long stay in such shelters. Huge areas of land, remote from the target zones, would remain uninhabitable for long periods, probably years if nuclear reactors were targets of attack, and most of the livestock and crops would be lost.

The effects of local fall-out would be felt just as badly in some non-combatant countries, but global fall-out would result in long-term damage in *all* countries; it would be expressed in an increased incidence of cancers and of genetic defects in future generations. The property of fall-out to extend the injurious action both in space and in time, is a novel and unique characteristic of nuclear warfare. Not only the inhabitants of the combatant countries, but virtually the whole population of the world, and their descendants, would be victims of a nuclear war—therein lies the radical change which nuclear weapons introduce into the whole concept of warfare.

Recognizing the ubiquitous nature of the harmful action of ionizing radiations, the United States and the Soviet Union had jointly proposed the banning of radiological weapons. However, this proposal did not include a ban on radiation resulting from the explosion of nuclear weapons, because it is impossible to produce a nuclear explosion without the emission of ionizing radiations. But if the use of radiation in warfare is considered to be unacceptable by civilized standards, then the only logical alternative is to ban nuclear war altogether. In any case, the ever decreasing security of the countries engaged in the nuclear arms race should convince their governments of the folly of continuing that race, and of the urgent need to limit and eventually eliminate all nuclear weapons.

Nuclear war, which many years ago was declared a crime against humanity by the United Nations, was again condemned by the General Assembly of the United Nations at the Special Session devoted to Disarmament, in 1978. The Final Document adopted by the entire United Nations membership includes the following warning: "Removing the threat of a world war—a nuclear war—is the most acute and urgent task of the present day. Mankind is confronted with a choice: we must halt the arms race and proceed to disarmament or face annihilation".

References

References are given only to material which does not appear in textbooks on the subject. The main text book used is Glasstone, S. and Dolan, P.J., *The Effects of Nuclear Weapons,* Third Edition (U.S. Departments of Defense and Energy, Washington, D.C., 1977). Material from this book was used in the preparation of tables 11, 15–17, 24 and 25, and for figures 2–4, 12–14 and 18–23.

1. Lapp, R.E., *Kill and Overkill* (Weidenfeld and Nicolson, London, 1962), p. 37.
2. Auxier, J.A., *Ichiban,* ERDA Critical Review Series (ERDA, Washington, D.C., 1977), p. 73.
3. *The Effects of Nuclear War* (Office of Technology Assessment of the Congress of the United States, Washington, D.C., 1979).
   (a) ———, pp. 122–23.
   (b) ———, pp. 94–106, 113.
4. *Comprehensive Study on Nuclear Weapons,* UN Report A/35/392, Report of the Secretary-General (United Nations, New York, 1980).
   (a) ———, p. 31.
   (b) ———, p. 69.
5. *The Effects on Populations of Exposure to Low Levels of Ionizing Radiation: 1980,* Report of the Committee on the Biological Effects of Ionizing Radiations (BEIR), (National Academy Press, Washington, D.C., 1980), p. 3.
6. Kerr, G.D., Organ dose estimates for the Japanese atomic-bomb-survivors, *Health Physics,* Vol, 37, 1979, pp. 487–508.
7. *Radiobiological Factors in Manned Space Flight,* ed, W.H. Langham (National Academy of Sciences, Washington, D.C., 1967), p. 77.
8. Tobias, C.A. and Todd, P., *Space Radiation Biology and Related Topics* (Academic Press, New York, 1974).
   (a) ———, p. 487.
   (b) ———, p. 497.
   (c) ———, p. 501.
9. Ash, P., 'The influence of radiation on fertility in man', *British Journal of Radiology,* Vol. 53, 1980, pp. 271–78.
10. Ichimaru, M., Ishimaru, T., Mikami, M., Yamada, Y. and Okhita, T. (personal communication 1980).
11. Loewe, W.E. and Mendelsohn, E., *Revised Estimates of Dose at Hiroshima and Nagasaki and Possible Consequences for Radiation-Induced Leukaemia,* Report D 80/14, Lawrence Livermore Laboratory, 1980.
12. Mancuso, T.F., Stewart, A. and Kneale, G., 'Radiation exposures of Hanford workers dying from cancer and other causes', *Health Physics,* Vol. 33, 1977, p. 369.
13. Arnold, J.R., 'The Hydrogen-cobalt bomb', *Bulletin of the Atomic Scientists,* Vol. VI, 1950, p. 290.
14. Watanabe, S., *Cancer and Leukaemia Developing Among Atom Bomb Survivors,* Handbuch der allgemeiner Pathologie, vol. 6, part 5 (Springer-Verlag, Berlin, 1974).
15. Johnson, G., 'Paradise lost', *Bulletin of the Atomic Scientists,* Vol. 36, No. 10, 1980, p. 24.
16. Peterson, K.R., 'An empirical model for estimating world-wide deposition from atmospheric nuclear explosions', *Health Physics,* Vol. 18, 1970, pp. 357–78.

17. Krey, P.W. and Krajewski, B., 'Comparison of atmospheric transport model calculations with observations of radioactive debris', *Journal of Geophysics Research,* Vol. 75, 1970, p. 2901.
18. Long-Term Worldwide Effects of Multiple Nuclear-Weapons Detonations (National Academy of Sciences, Washington, D.C., 1975), p. 105.
19. Braland, W., Franke, B. and Teufel, D., *Transfer of Organically Bound Radionuclides Through Food Chains to Man,* IAEA-SM-237-17, Vienna, 1979.
20. *Warfare in a Fragile World: Military Impact on the Human Environment* (Taylor & Francis, London, 1980, Stockholm International Peace Research Institute).
21. Sasser, I.B., *Fall-Out Radiation Effects on Livestock and Food Crops* (Oak Ridge, Tennessee, 1971).
22. *The Physical and Medical Effects of the Hiroshima and Nagasaki bombs,* International Symposium on the Damage and After Effects of the Atomic Bombing of Hiroshima and Nagasaki, Hiroshima, 1977.
23. Bennett, B., *Fatality Uncertainties in Limited Nuclear War,* Report Rand/R-2218-AF (Rand Corporation, Santa Monica, 1977).
24. *The Effects of Nuclear War* (U.S. Arms Control and Disarmament Agency, Washington, D.C., 1979), p. 20.
25. Shute, N., *On the Beach* (Heinemann, London, 1957).
26. *Protect and Survive* (HMSO, London, May 1980).
27. Sidel, V.W., Geiger, H.J. and Lown, B., 'The physician's role in the postattack period', *New England Journal of Medicine,* Vol. 266, 1962, pp. 1137–45.
28. *Arms Limitations, Security and Development,* Proceedings of 30th Pugwash Conference, Breukelen, 1980, p. 21.
29. Fetter, S. and Tsipis, K., *Catastrophic Nuclear Radiation Releases,* M.I.T. Report No. 5 (MIT, Cambridge, Mass., 1980).
30. Committee on Disarmament document CD/31, 1979 (Joint USSR-United States proposal on major elements of a treaty prohibiting the development, production, stockpiling and use of radiological weapons).
31. Wilrich, M. and Taylor, T.B., *Nuclear Theft: Risks and Safeguards* (Ballinger, Cambridge, Mass., 1974), p. 20.
32. *Nuclear Proliferation and Safeguards* (Office of Technology Assessment, Washington, D.C., 1977), p. 140.

Index